The Well From Hell

Also by William Sargent:

Sea Level Rising, 2009

Just Seconds from the Ocean, 2008

Crab Wars, 2006

The House on Ipswich Marsh, 2005

Storm Surge, 2004

A Year in The Notch, 2001

Praise for Other Books:

"With his fine descriptions and lucid explanations, Sargent joins the company of Lewis Thomas and Stephen Jay Gould as a first-rate interpreter of modern science."
~ Publisher's Weekly

"It is a gem of Natural History... the best introduction to the original environment of the New England coast."
~ Dr. E. O. Wilson,
Harvard University

"A joy to read."
~ The Washington Post

"A Great Read! Sargent takes us on a raucous jaunt through the New England forest, to see the big picture with unclouded eyes. A true biologist, he examines everything in sight and counts it relevant, connecting it with seamless prose into the rational
new picture. It's a powerful boost to the new Nature religion that references us to Life on Earth."
~ Dr. Bernd Heinrich

It's science writing that reads like a novel, with all the page-turning excitement of a thriller."
~ William Martin

"Sargent can turn an event as mundane as a rising tide into poetry. This is a book for everyone who loves the shore, especially Cape Cod."
~ The Boston Globe

"If you only have time for one book about life-death dramas played to the sound of crashing waves, about new science and the old sea, about Nobel prizes, squid brains and sex orgies on Cape Cod beaches, then this book is for you."
~ Dr. A.A. Moscona,
Journal of the American Medical Association

THE WELL FROM HELL

WILLIAM SARGENT

Strawberry Hill Press

ISBN-13: 978-1460972427

ISBN-10: 1460972422

Contents:

Preface
February 14, 2011

It is a warm, sunny Valentine's Day in mid-February. New England is digging out from one of the snowiest winters on record. Local papers are showing a photo of snow piled up against Shaquille O'Neal, the seven-foot tall 325-pound center of the Boston Celtics. Shaq has about 30 inches to go before he needs a snorkel.

The roofs of over 80 buildings have collapsed under the heavy snow injuring both cows and people. Editorials are suggesting that building codes should be changed to accommodate our changing local climate. Hundred-year-storms now occur every ten years.

The price of oil is settling back down after the Egyptian revolution, as are fears that strikes could close down the Suez Canal to the shipment of gas and petroleum. One of the many reasons that Mubarak had been overthrown was that the price of wheat, corn and soybeans had risen because of the effects of global warming.

These are all reminders of how much human civilization depends on oil, and the severe costs of that dependence. The BP spill lies at the nexus of all these problems. It is difficult to remember now, but the Deepwater Horizon oil rig exploded on exactly the same day that planes were allowed to fly again after having been grounded by an eruption of the Eyjafjallajokull volcano in Iceland. I remember it well because I was busy writing a book about Eyjafjallajokull, thinking that it was going to be the dominant environmental story of 2010. The reader may want to stop here if she is looking for accurate prognostication.

Many people were instrumental in helping complete this book. Jill Buchanan has been an indispensable publisher, cover designer, editor and savvy business manager. She did a masterful job editing the manuscript and designing the cover.

Rick Camilli and Chris Reddy do excellent research and were extremely useful in explaining the intricacies of chemical oceanography, but could use some help returning phone calls. Of course, they were pretty busy appearing before Congress and talking to President Obama, but the Woods Hole Oceanographic Institution and MIT were helpful in tracking them down. The Coast Guard and Thad Allen were the true heroes of both the oil spill and Katrina and helped explain some of the background of the spill.

On one of my trips to Louisiana I was supposed to be taken by boat out to see the cleanup operation, but the boat had been cancelled at the last minute. It looked like the entire trip was going to be a bust, until I happened to notice a young man leaning against a government truck in the parking lot of the small hotel where I was staying. On a whim, I started talking to him. It turned out that Michael Bey was an ecologist with the US Geological Survey and he asked me if I would like to join him collecting marsh data the following day. It also turned out that Michael Bey was about the best person in all of Louisiana to lead me through the marsh. His brother also runs a delicious sausage restaurant in New Orleans.

Finally, I would like to thank Kristina, Chappell and Ben. They have been with me through all the ups and downs of writing. It has not always been easy and I could never have done it without your love, understanding and support. Thank you!

Chapter One
Paleozoic Problems

Millions of years ago, our planet was in a global warming crisis. Runaway volcanoes were belching gigatons of heat-trapping carbon dioxide into the atmosphere and giant conifer forests towered over fetid swamps. The forests' immense trees swept carbon dioxide out of the air, then died and toppled back into the swamps, where their carbon-based trunks were sequestered under thick layers of anaerobic mud. Today, peat from those Carboniferous swamps make up our modern coal. But these early forests were limited in the amount of carbon they could sequester because they grew on land that only made up less than a third of the earth's surface.

There were also large, stone-like structures in shallow waters along the coasts of the ancient continents. These structures, called stromatolites, contained layers of bluish-green bacteria that sequestered carbon by pulling carbon dioxide out of the supersaturated seawater. The stromatolites' bacteria then mixed the carbon with calcium to construct their stony skeletons, much like modern corals do today. But, like the forests, the stromatolites were limited in the amount of carbon they could sequester because they could only live in sunlit waters less than a few feet deep.

Over millions of years, however, the forests and stromatolites removed enough carbon dioxide to cool our planet and stop the runaway greenhouse effect, but this shifted the earth into a bone-chilling ice age. Glaciers formed and sea levels dropped, killing the stromatolites as they became stranded on dry land. Without the stromatolites to sequester carbon, the earth once again careened back into a period of global warming. Our planet remained locked in this cycle of swings from one extreme to the other for hundreds of millions of years without a mechanism to temper the drastic changes.

About 160 million years ago, something mysterious and quite wonderful happened. A tiny new creature evolved that could draw carbon dioxide out of seawater and use it to construct its intricately beautiful calcareous shells. What was truly revolutionary about this shell-forming sea creature, was that it also evolved a planktonic lifestyle. Unlike the static stromatolites that were forever tied to the shallow waters on the edge of continents, these new creatures could drift freely over the world's oceans that made up 70 percent of our planet's surface and almost 90 percent of its volume. That original species then evolved into today's riotous collection of gorgeously beautiful planktonic creatures with equally beautiful names; the foraminifera, globigerina, coccolithophores and "wing-footed" pteropods.

Billions of these tiny, planktonic calcifiers grew and reproduced. They could exist anywhere in the world's oceans but they thrived best over deep-water basins where upwelling currents bathed them in a constant supply of nutrient-rich waters that fertilized their growth.

When these plankton died their calcareous shells rained down onto the deep ocean floor. The tiny droplets of waxy oil the plankton had used in life to keep them afloat accumulated into thousand-foot-thick deposits of calcareous ooze. Over time, these deposits were buried beneath more deep layers of sediments, and they started to cook in the heat from the earth's interior.

Since phytoplankton is so ubiquitous, you would think oil would underlie all the world's oceans. But petroleum is volatile stuff. If the temperature remains too cool the ooze will never cook; too hot, and the ooze will boil off as natural gas and quietly dissipate. However, if the temperature stays just right, between 212 and 275 degrees Fahrenheit, for just the right amount of time, about a million years or so, the oil simmers into a golden-brown rue of sweet, low-sulfur crude.

Only relatively small pockets of oil end up heating, under just the right conditions, for just the right amounts of time, to form commercially viable quantities of petroleum. One of the areas where this happened was below the ancient seabed of the proto Gulf of Mexico.

About 165 million years ago, the Yucatan Peninsula started to rift away from the North American plate. In its wake it left a restricted basin of warm, shallow water. Seawater would occasionally slosh into this basin and evaporate, leaving behind a thick layer of salt. The salt would be buried under succeeding layers of salt so that today the entire basin of the Gulf of Mexico is underlain by a five-mile-thick foundation of salt, called the Louann Formation.

Several ancient rivers also drained the early American continent, as the Mississippi River does today. At the same time, plate tectonics drove the American plate over a hot spot of rising magma, pushing up the earth to create the Rocky Mountains, raising the continent's interior, and finally creating the volcano that made the modern-day island of Bermuda.

About 20 million years ago, during the Miocene, the Appalachian Mountains were pushed up by plate tectonics, creating fast flowing rivers that carried pulses of sand and mud to the eastern side of the Gulf of Mexico. Immense underwater fans of sand spread out over the slippery Louann Salt Foundation to fill up the basin of the Gulf of Mexico. The mud and sand formed river deltas and offshore islands much like those off the Mississippi River today.

As the mud and sand accumulated, they became compressed and eventually turned into sandstone and shale. Below all of this was the oil that, over time, migrated up through the formations, pooling in the porous sandstone and getting trapped beneath the impermeable strata of shale.

The shale and sandstone slid slowly over the salt foundation toward the depressed center of the Gulf of Mexico. The sliding caused the sandstone to sink down through the salt to form pockets and basins of oil, much like rock salt melts into snow. At the same time, pressure from the overlying rock caused the salt to squeeze up through and around the sandstone deposits in huge towering columns and overarching domes. The salt domes helped protect the nascent oil from overheating.

This then, was origin of sweet Louisiana crude, the genie in the lamp that saved our planet from global warming hundreds of millions of years ago. Today, that genie can either power our civilization for another 50 years, or plunge us back into another crisis of runaway global warming.

Since the early 1900's, our species has created the planet's four largest corporations to find and release these genies from their oily lamps. The genies have been found in Pennsylvania, California, Russia, in the ancient kingdom of Persia, off the deep-sea coasts of Brazil, the North Sea, and in the Macondo oil field in the Gulf of Mexico. Now, we have to decide if we want to put those genies back in their oily little subsea bottles once again.

Chapter Two
D'Arcy's Legacy
Masjid-i-Suleiman, Persia
May 25, 1908

The sun shone down relentlessly on the sere foothills of Masjid-i-Suleiman. For the past four days, temperatures had topped 110 degrees. Diaphanous ghosts of vapor shimmered above an open drill hole, and the smell of natural gas hung in the heavy night air. It was the smell that excited oilmen, the smell you longed for when drilling a wildcat well into an unexplored oil field.

Lieutenant Arnold was exhausted from the pressure and heat of the past few months. His men had spent almost a week trying to retrieve a drill bit that had become unscrewed while drilling through the hardest rock they had yet encountered. They had also decided to ignore a cable from their financiers ordering them to cease operations on the dry hole. What did a bunch of Scottish merchants know about the smell of methane?

It was so hot that Lieutenant Arnold was sleeping outside his tent. But at 4:00 a.m. he awoke with a start. The ground was shaking and his men were yelling. He raced to the derrick site. A towering geyser of rich black oil was gushing 50 feet over the top of their rig. It threatened to smother his workers in oil, if it didn't suffocate them in the miasma of methane now billowing over the parched landscape.

William Knox D'Arcy was entertaining in London when an aide handed him the cable, "See psalm 104, verse 15, third sentence." D'Arcy didn't need a bible. Oilmen knew the relevant passage by heart, "…that he may bring out of the earth oil to make a cheerful countenance."

D'Arcy had won another long shot. He smiled at his wife, but held counsel from his guests. He had always loved a bet. First, it was the horses he developed a fondness for when he emigrated to Australia, then it was the defunct gold mine he brought back into production so profitably.

It was the Australian gold that had allowed D'Arcy to return to England as a wealthy man. Now, he owned a townhouse in Grosvenor Square, two country estates, and boasted a stunning second wife—the former actress Nina Boucicault, who often invited her friend Enrico Curuso to sing at the D'Arcy's lavish private parties. Besides the queen, they were the only people in England who owned a private box at the Epsom racetrack.

However, D'Arcy never lost his lust for taking yet another chance. When he heard of an obscure geological report that indicated there might be hidden reserves of oil beneath the deserts of Persia he had jumped at the opportunity.

It had not been difficult to win the concession from the equally profligate Shah, who was continually broke from his own frequent visits to European spas. The Shah's latest project was to introduce the new, French filmmaking technology into Persia so his servants could watch the risqué foreign films.

D'Arcy offered the Shah 5,000 pounds to get things started, then an additional 20,000 pounds upon signing and 20,000 pounds worth of shares in the Anglo-Persia Oil Company. Persia would receive only 16 percent of the company's annual profits. In return, D'Arcy could extract oil from three quarters of all the land in Persia, an area twice the size of the oil-soaked state of Texas.

Marcus Samuel

While D'Arcy was the colorful gambler always on the hunt for new sources of petroleum, Marcus Samuel represented the opposite pole of the British oil industry. He was the careful businessman, ever ready to negotiate the best price for his foreign oil.

Samuel had grown up in the East End of London, where his father owned an import business that sold shells to wealthy Victorian ladies.

After he inherited his father's company, Samuel quickly realized that shipping was key to any importing business, and that oil was the key to shipping. He became an oil merchant but named his new company Shell to honor his father, and gave his oil tankers the obscure names of mollusks like Murex, Pectin or, simply, Clam. But the names were significant, because Samuel put as much thought into his ships as he did into his cargo.

The British oil industry needed both D'Arcy and Samuel, because Great Britain had no oil of it's own. It was entirely dependent on finding foreign oil and paying the cheapest prices to import it to Britain's isolated island ports.

The English have always regarded oil differently than have the Americans. Oil was once so plentiful in the United States that it seeped naturally out of the ground in places like Pennsylvania, Texas and even at the La Brea tar pits in downtown Los Angeles. L.A. residents still get tax benefits from the oil that is pumped from beneath their city streets. Up until the 1910's, the United States produced between 60 and 70 percent of the world's oil supply. Britain, on the other hand, was the leader in global exploration, and by 1919, controlled 50 percent of the world's proven oil reserves.

Another example of how both countries think about oil differently is reflected in the way they talk about petroleum prices. In the United States, we still price oil in terms of dollars per barrel, because it was originally shipped in the same wooden barrels as those made for Pennsylvanian whiskey. The English price their oil in pounds per ton, because that is how petroleum arrives in their ports.

Far-sighted British oilmen realized they were going to have to both find oil in often unstable parts of the world and ship it back profitably to England. They started to think of oil as a rare and uniquely valuable commodity, one destined to become forever linked to empire, diplomacy, even survival itself. This was in marked contrast to the "drill, baby, drill" philosophy most-recently articulated by the gun-toting former governor of Alaska, home to some of the largest and most fragile supplies of oil on our planet.

Wits and Water

For years, Marcus Samuel had been trying to convince the British Royal Navy to switch from burning coal to burning oil. He argued that ships could travel four times faster and four times farther using the same amount of oil as coal, and you didn't need so tie up so much manpower in the unending task of shoveling tons of coal into seemingly bottomless burners.

But the Admiralty was leery of change. After all, Britain had its own abundant supplies of native coal, so why switch to a scarce commodity that could be easily monopolized by foreigners like the Russians or that crafty Yank, Rockefeller?

One rising young British politician was slightly less blinkered by such parochial prejudice than his countrymen. Winston Churchill had assumed command of the Admiralty in 1911, and was already convinced that Marcus Samuel was right. But Churchill didn't quite trust Shell Oil. After all, the rest of the world knew his company as Royal Dutch/Shell Oil. It was half owned by the Netherlands, and the Netherlands were close to Germany, which Churchill already regarded as Great Britain's future enemy.

Besides, Churchill also seemed to just prefer the gambling, colorful, risk-taking pole of the British oil industry represented by Knox D'Arcy over the businesslike pole of the industry represented by Marcus Samuel.

It is also reputed that D'Arcy had already paid Churchill 5,000 pounds to win the Persian concession, and that Churchill trusted a syndicate of patriots, including D'Arcy and his backers, the Scottish merchants of Burmah Oil, rather the careful businessman, Samuel, with his ties to the Netherlands.

Consequently, in 1914, Winston Churchill rose before Parliament and used his considerable rhetorical skills to convince its members to purchase 51 percent of the Anglo-Persia Oil Company. It was an almost unprecedented step. The only other time England had bought into a private company was when it took over control of the Suez Canal to defend its withering empire.

Parliament had essentially nationalized Great Britain's oil industry while obtaining a secure supply of Persian oil just prior to the outbreak of World War I. But England had also just earned the enmity of generations of Iranians who seethed under D'Arcy's lopsided deal with the former Shah.

This was also the beginning of the intertwined relationship between the British government and what would become British Petroleum. Government officials and company executives were cut from the same cloth. They tended to be products of Eton, Oxford and Cambridge and belonged to the same gentlemen's clubs.

British Petroleum executives grew to expect and receive deference from the British government. This was in marked contrast to the United States, where the Washington establishment had traditionally held the oilmen at arm's length. After all, the hugely popular trust buster, Teddy Roosevelt, had taken on the equally unpopular monopolist John D. Rockefeller and successfully broken up Standard Oil into seven smaller, competing companies.

An American oilman once enviously noted that, "In England they knight their successful businessmen; in America we indict ours."

This expectation of deference would not serve BP's leaders well when they had to testify before the United States Congress in 2010.

Check Mate

The Anglo-Iranian Oil Company held a monopoly on Iranian oil throughout most of the First and Second World Wars. However, in 1941, two months after Germany invaded the Soviet Union, British and Russian forces moved into Iran to protect the Abadan refinery. The allies then deposed the Reza Shah, whom they suspected was sympathetic to the Nazis, and replaced him with a new Shah, his 21-year-old son, Mohammed Pahlavi.

Pahlavi was beset by warring factions throughout his reign. On one side were the Islamist fundamentalists led by Ayatollah Seyed Kashani, and on the other side were the communists led by the Tudeh party. The only thing that united all of Iran was its hatred for the Anglo-Iranian Oil Company, led by Sir William Fraser, an old-school Scotsman who refused to budge from ever giving Iran anything more than the piddling 16 percent on the profits from its own oil.

Things finally came to a head when the Iranian parliament nationalized the oil industry and elected Mohammed Mossadegh as their new prime minister. A sheep was sacrificed in front of the headquarters of Anglo-Iranian Oil, and the crowds were told that from now on all of Anglo-Iranian's oil and assets were the property of Iran.

By 1953, the United States was so concerned that Mossadegh was leaning toward the Soviet Union that it hatched out Project Ajax, a CIA plot to stage a coup and have the Shah dismiss Mossadegh. But "Old Mossy" figured out the caper and instead had his supporters flood into the streets shouting against the Shah's perfidity. The Shah fled to Italy, only to return immediately after the head of the army, General Zahedi, assembled enough imperial troops to stage a countercoup. The duly elected Mossadegh was put back into prison, where he had spent so much time under the reign of the Shah's father.

In the end, the Anglo-Iranian Oil Company lost it's monopoly to a consortium of its competitors, the five largest other oil companies in the world. They eventually had to pay Anglo-Iranian for its losses. John Louden, the manager of Anglo-Iranian's old rival Royal Dutch/Shell, pointed out that, "It was the best deal Willie Fraser ever made," because BP actually didn't have anything to sell, since the Iranian oil industry already been nationalized.

Arguably, it was also the best thing that ever could have happened to Anglo-Iranian. Now, instead of being a quasi-government corporation with a built-in monopoly on Iranian oil, it would have to go out and compete with the Yanks to discover new sources of oil. In the process, it would become the second largest oil company in the world, with oil fields around the world, including in the backyards of its major

U.S. competitors. To do so, it would have to take some risks, perhaps even cut some corners, but that, too, was part of the legacy handed down by the inestimable Mr. D'Arcy.

Chapter Three
The Sun King

"There are two things you don't want in BP. First is to work for John Browne; second is to have John Browne work for you."

~ David Simon,
Chairman of BP's Board of Director, 1997

During the 1960's, British Petroleum was primarily concerned with finding oil to replace the concession it had lost in Iran. The quest led the company to Kuwait and then Libya, but it would lose both these deals to nationalization, as it had in Persia.

The company could very well have gone bankrupt during the 1973 oil embargo had it not been for its 1965 discovery of over 30 years worth of domestic oil off the coast of Aberdeen, Scotland, and its 1969 discovery of oil in Prudhoe Bay, Alaska.

British Petroleum was initially thwarted from moving its Alaskan oil to market by environmentalists who were concerned that a pipeline carrying warm oil would melt the permafrost and disrupt the paths used by migratory caribou. Such concerns didn't stand a chance after the embargo raised the stakes for obtaining supplies of domestic oil.

Another environmental concern was also emerging during this period. Geologists were starting to realize that the demand for oil would eventually outstrip industry's ability to discover new resources. This was a radically new concept, given that oil companies' main concern until then had been how to avoid creating gluts of new oil that would drive down profits.

Today, of course, the term "peak oil" is fairly commonplace in referring to the time when the reserves of oil will peak and prices will start to climb as the supplies of oil decline. Some economists think we hit peak oil around 2006, but most oil executives hope that drilling for more and more oil in deeper and deeper waters can avert the peak for a few more decades.

Back in the early '70's, scientists were also seeing presentations on the first measurements of atmospheric carbon dioxide, which were made by Charles Keeling at the Mauna Loa Observatory in Hawaii. Al Gore has famously written about seeing the iconic saw-toothed graph in an undergraduate seminar at Harvard College.

The graph shows our planet breathing. In the spring, the jagged red line falls as northern forests inhale carbon dioxide through their new, green leaves; in the winter, the line rises as leaves decompose and forests exhale carbon dioxide back into the atmosphere.

But the graph also shows another trend. The saw-toothed line rises inexorably to the right. That is because each year that the earth exhales, it releases more carbon dioxide than the year before. This annual increase represents the amount of carbon dioxide added to the atmosphere from the burning of fossil fuels.

The graph had a galvanizing effect. Scientists instantly recognized the chart as illustrating how the increase in carbon dioxide would inevitably lead to the greenhouse effect, glacial melting, summer heat waves, sea level rise, and all the other manifestations of global warming we are witnessing today. It was only later that oil company executives and industry-supported scientists would try to debunk the new information.

Even as callow undergraduates, Al Gore's classmates could grasp the implications of the simple graph. But most assumed that since scientists already knew about the problem, it would be solved by the time they graduated from college. That was over 40 years ago. What have we done about global warming since then? Virtually nothing.

In the interest of disclosure, I admit that I was one of those undergraduates sitting beside Al Gore in that seminar. However, unlike Gore, who was taking the seminar for credit, I was a simple auditor. In the interests of *full* disclosure, I will further admit that I am still ticked at Harvard for turning me down for the prestigious seminar.

Another undergraduate science major was also being exposed to these discoveries at Cambridge University in England. John Browne was well on his way to a quiet life, studying physics in academia. But Browne's father had convinced John to apply for a summer internship in British Petroleum's exploratory division in Anchorage, Alaska.

Browne soon discovered that he was not only fascinated with America, but with the rough-and-tumble world of oil exploration. In lieu of pursuing his academic career at Cambridge University, he stayed with British Petroleum in Alaska, where he became involved in many of the company's early environmental debates and held a variety of posts. Perhaps he even noticed that Alaska's climate was changing and her famed glaciers were starting to melt.

Browne decided to ditch the pursuit of a PhD and continue his education at Stanford University's School of Business, where he discovered he was just as intrigued by the financial side of the oil industry as he had been with the exploratory side. As he continued working for British Petroleum throughout the United States, Browne started to appreciate America's more democratic and aggressive way of doing business.

He also realized that if he played his cards right, he might some day end up in a position not only do something about the culture of British Petroleum, but the culture of the oil industry itself. His time would quickly come.

In 1982, Browne, as a still-young employee, devised a clever financial scheme that had significant implications for the future of British Petroleum. If the company sold .25 percent of its North Sea holdings to rival British firms, it could not only drastically reduce its taxes, but also make a tidy $300 million profit. Now that was the sort of idea that garnered you instant attention in the oil business.

British Petroleum had recently purchased 55 percent ownership in Sohio, the former Standard Oil of Ohio. The deal included Robert Horton, a fellow Brit and MIT-trained director of Sohio, who identified a kindred spirit in Browne. He realized that if they teamed up they could jettison British Petroleum's stodgy, old-school way of doing business in favor of the new, more free-wheeling, American model. In 1986, Horton made Browne chief financial officer of Sohio. Together, they would make history.

At the time, Jack Golden was one of British Petroleum's senior scientists. He had always harbored a geologist's hunch that vast deposits of oil lie in the deep waters just beyond the continental shelf of the Gulf of Mexico—waters far deeper than where any the oil companies had drilled before.

When Golden showed his seismic data to Browne, he could see that the new CEO instantly understood the implications. Browne sank Sohio's entire $55 million exploratory budget into the new site. The gamble paid off. By 1989, British Petroleum controlled over a third of all the oil fields in the Gulf of Mexico and was the largest oil producer in the United States.

Browne's bet had been as audacious as any made by William Knox D'Arcy. And, like D'Arcy's, it paid off handsomely. Although the initial cost of exploring deep-sea areas was steep, once the oil was discovered, it only cost $6 per barrel to pump it ashore, where it would fetch five times that amount on the market.

Browne's gamble had also set the future course for British Petroleum. Throughout the 1990's the company would concentrate on leasing expensive, new, semi-submersible deep-sea rigs to drill for the next big find, another offshore "elephant" to forestall peak oil and support the company for years to come.

Beyond his high-risk, profitable investments, John Browne's swashbuckling style cemented his reputation within the company. British Petroleum's board of directors liked the sort of chap who harkened back to D'Arcy's halcyon days. The Board made Browne CEO of the entire British Petroleum group in 1995. Browne was finally in a position to be able to shake up the comfortable world of the oil industry. He decided to do it in a grand manner.

He chose his old alma mater, Stanford University, as the stage. The School of Business had invited him to give a lecture to a select audience, including several fellow oil executives.

He opened his 1997 talk by admitting that he had been mistaken in his previous skepticism toward global warming. The evidence was clear: burning fossil fuels was contributing to global warming, and oil companies were in a unique position to do something about it. Murmurs of horror rippled through the hushed auditorium. Browne had broken ranks in a most undignified and public manner.

Browne made good on his speech. He pledged to reduce British Petroleum's output of carbon dioxide by 10 percent by the year 2010, and then surpassed that goal eight years early. He spent $160 million installing solar panels on British Petroleum's buildings in California and Spain and withdrew from the Global Climate Coalition, an oil industry-sponsored group established to refute the scientific claims that burning fossil fuels was causing global warming.

Then, after months of discussions with the international marketing and PR firm Ogilvy and Mather, Browne launched a $200 million campaign to re-brand British Petroleum as simply "bp" which stood for "beyond petroleum." He was delighted with the advertising agency's stunning design for the company's new logo. It was the sun goddess, Helios, surrounded by a burst of green-and-yellow sunflower petals. The lower case "bp" wouldn't stand the test of time, but the reinvented company BP had been born.

Browne helped draft a new mission statement, promising that BP would reinvent itself as "an energy company that people could have faith in."

BP had its mojo back. Morale soared and the media dubbed Browne "The Sun King." He was voted Great Britain's most admired businessman in 2001, 2002 and 2003.

Environmentalists would later argue that Browne's entire campaign had been a hypocritical move to greenwash the company. One environmental organization even awarded BP an emerald paintbrush for its media campaign. But the harshest criticism came from executives of rival oil companies who were envious that Browne was receiving so much attention for beating them to the punch.

Yet, evidence shows that Browne was, in fact, sincere in his efforts, that he was ahead of his time, and certainly ahead of the oil industry, to say nothing of his own board of directors.

Ultimately, however, Browne was a cost cutter, and his interest in profit trumped concerns for global warming and human safety. The Stanford-trained businessman coolly replaced BP's cadres of well-trained engineers with cheap outside consultants and engineers-for-hire, like Halliburton and Schlumberger, who could be had for a fraction of the cost of regular employees.

The Year of the Accidents

In March 2005, BP's Texas City oil refinery exploded in a massive fireball that killed 15 workers and left 170 more injured. Billowing clouds of acrid smoke trapped neighbors in their stifling, oil-suffused homes for days on end.

BP was eventually fined $21 million for safety violations leading up to the explosion. It had been a horrible accident, but apparently not horrific enough to have unduly interrupted Browne's holiday with his good friend Jeff Chevalier in their private apartment in the Doge section of Venice.

A year later, Browne was at a board meeting at the Intel Corporation when he received word that workers had discovered oil leaking along a 16-mile section of BP's shared pipeline in Prudhoe Bay, Alaska. As it turned out, the pipe, which hadn't been cleaned in over a decade, had built up excessive amounts of sediment and was corroded. As much as 267,000 gallons of oil spilled over 1.9 acres, creating the largest Alaskan oil spill in history. The pipeline had to be shut down for several months. It was another serious black eye for BP, but apparently also not quite serious enough for Browne to have interrupted the board meeting.

BP's reckless cost-cutting practices continued to play out in July when Hurricane Dennis swept through the Gulf of Mexico. After the storm, a passing boat noticed BP's oil platform, Thunder Horse, listing to one side. The giant rig, which towered 15 stories above the water's surface, was intended to produce about 20 percent of the Gulf's oil output. As it turned out, a valve had been installed backwards, causing the rig to swamp and almost plunge to the bottom. Further investigation revealed other shoddy work, including welding that left underwater pipelines brittle and full of cracks.

Again, outside consultants were found to be the cause of the accident. One of the "rent-an-engineer" firms had recommended that cheap metal valves be used instead of the standard-issue nickel valves that were specifically manufactured to withstand pressures encountered in the deep-water environment.

Timing of the accident was particularly embarrassing for Browne, who had been en route to a formal dinner at England's Greenwich Maritime Museum. Awkward, yes, but again, apparently not quite awkward enough for him to leave the evening's festivities early.

After the accident, BP started leasing platforms from outside companies like Transocean, an American firm that would move its headquarters first to the Cayman Islands, then Switzerland, to avoid paying United States taxes.

The embarrassments continued throughout the year as hurricanes Katrina and Rita halted drilling in the Gulf of Mexico. But, by this time, BP's board of directors were starting to get the idea that Browne chafed at the day-to-day drudgery of running an oil company. He seemed to have a narcissistic personality, only interested in pursuing the next big deal, whether it be buying the American giants Amoco and Arco, which had made BP the second largest oil company in the world, or, later trying to buy Shell, BP's old rival from the other side of the tracks in London.

Just a few years back, the media had dubbed Browne the "Sun King," but now his detractors started referring to him as

"Saint John the Divine." He ignored his critics, focusing instead on professional and personal pursuits, among them, building up his own impressive private collection of Romantic-era Italian paintings and pre-Columbian artifacts, and eventually becoming a director of the Tate contemporary art museum in England.

Throughout his career, however, John Browne had harbored a liability. He was gay. Most of the time it was not a problem. He lived in fashionable London and made a point of always forthrightly informing his board of directors of his orientation.

But when his mother died in 2001, and later when Jeff Chevalier returned to Canada, Browne seemed to lose his bearings. He had always been close to his mother, who had survived Auschwitz. They had lived quietly together in Chelsea, England, until she died. After she died, Browne started going out more openly to enjoy London as a billionaire gay bachelor.

His job at BP had always been the anchor that grounded his workaholic personality. But Browne was up for retirement at age 60, and BP's board of directors was ready for a change.

Browne decided to forestall his retirement by making himself indispensable to BP. He attempted to merge BP with its old rival, Shell. The merger would have made BP the largest corporation on the planet and John Browne indispensable at running the behemoth.

It was too much for BP's Board of Directors to swallow. This time Browne had gone too far. They felt he was starting to harbor delusions of grandeur and had become addicted to the spotlight.

However, Browne had another problem, which the board was unaware of. Since 2006, he had been battling with a London paper to block publication of his messy break-up with Jeff

Chevalier, whom Browne had met through a gay escort website called "Suited and Booted." Browne was trying to win an injunction to block the embarrassing disclosures at the same time he was fighting for his job.

He started a media campaign that he thought might shore up his position with the board. The *Financial Times* ran a flattering portrait of his activities that headlined the paper for an unprecedented three days.

But the puff piece only made things worse. The chairman of BP's board of directors, Peter Sutherland, expostulated publicly about the "madness" of such American-style self-aggrandizement.

Sutherland's warning shot did not deter Browne from escalating the argument further. He allowed Merrill Lynch to circulate a note to its clients that described Browne's retirement as a medium-term risk for investors.

It seemed that Browne had become so infatuated with his own importance that he felt he could use such outside pressure to influence his board. Little did he know that the board had already met privately in Vienna and agreed that he would have to leave by the summer of 2008, the 100th anniversary of D'Arcy's founding oil strike in Persia.

The chairman summoned Browne into his office and gave him the news. Browne said he would think about it. But on May 1, 2007 he changed his tune. Over a breakfast of marmite and scones, he told Sutherland that he had failed to keep Chevalier's lurid revelations out of the tabloids. What's more, a judge had ruled against Browne, charging that he had lied to the court by saying he had met Jeff Chevalier exercising in Battersea Park rather than online.

Browne still hoped he could remain at BP to preserve a

shred of dignity, but later that morning he met with two trusted media specialists who advised him that his position was no longer tenable. After loyally working for 41 years for a company that had given his life so much meaning, Browne resigned.

The board decided to replace Browne with a family man who had a solid background in engineering. Someone who wasn't always so concerned with what the media was going to say about his performance. Tony Hayward would be just the man to lead BP into the next century.

The Price of Oil

BP could have used its string of crises in 2005 and John Browne's resignation as catalysts to initiate a top down review of the company's safety and environmental procedures. Exxon did this after the Exxon Valdez accident and came out of the process with a set of stringent operating procedures that made it a leader in the field.

Unfortunately, while Tony Hayward said all the right things, promising "a fundamental shift in the way BP works," nothing ever seemed to change. His attention was drawn instead to BP's partnership with the Russian oil company Tyumen (TNK) that owned extensive oil fields in Siberia.

TNK was partly owned by BP and partly owned by a group of Russian oligarchs who had used their own sharp-elbowed tactics to gain control over Tyumen when it was privatized after the break-up of the Soviet Union.

But the oligarchs were a sensitive bunch. They were put off by what they perceived as Tony Hayward's condescending attitude toward Russians. As one oligarch put it, "I've read many more books and know a lot more about art and music than Hayward, so I don't know why he thinks he is so superior."

Hayward seemed oblivious to the impression he made on other people, and he certainly didn't expect any problems from the oligarchs at TNK-BP's annual meeting in Antibes. But before the meeting started, Mikhail Fridman, one of the oligarch owners, approached Hayward, "Tony, we have bolshoi problema. President Putin says we cannot have any non-Russians running a Russian company." Hayward shrugged, "Thank you, Michael, but I would have to hear that from the president, personally."

Things really started to unravel for BP in 2008, after Putin was made prime minister. BP's representative on TNK-BP was its long-time American executive, Bob Dudley. The oligarchs on the board resented Dudley and had quietly let his contract expire. This was something that often happened. Without a contract, Dudley's work visa could expire, forcing him to leave the country.

On March 19, 2008, none of TNK-BP's Russian employees showed up for work. Shortly after Dudley arrived, 50 officers from Russia's Interior Department barged into the office demanding to see all employee work contracts; then they proceeded to lock up all the file cabinets and set up cots for round-the-clock surveillance.

On July 24, Dudley received word that he would be arrested if he tried to stay in Moscow another day. He didn't even bother to pack, driving straight to the airport to catch the next flight to Paris.

Two days later, Hayward capitulated, signing a Memorandum Of Understanding with TNK that BP would find an independent Russian to run the joint company. The compromise allowed BP to retain its share of the profitable business, but BP still needed new oil.

Something else was roiling the oil industry in the early 2000's. Ever since the mid-1980's, the price of oil had hovered under $25 per barrel. But in 2003 it started to rise, and by July 2008 it peaked at $147.50 per barrel, with American gasoline prices topping an unheard of $4 per gallon.

The reasons for the price hikes were many. China had stockpiled diesel to prevent shortages during the Beijing Olympics. European drivers had started to switch from gasoline to less-expensive diesel. Traders bought extra supplies of diesel because European and Asian refineries were limited in the amount of diesel they could refine. The U.S. Department of Energy bought 30,000 barrels of oil to refill the U.S. strategic petroleum reserves, and U.S. refineries cut back on their production of diesel because of new limits on the amount of sulfur it could contain.

The chairman of the Libya Oil Company, Shokri Ghanem, saw the problem from a different angle, stating that OPEC countries were reaching their peak production of oil, "There is no more oil the countries can produce. We are squeezing the rocks as hard as we can."

Despite these problems, the main reason that oil prices rose so precipitously was because of speculation. There was just too much money available to invest and too many shiny new financial instruments to make it all happen. When oil prices rose to over $140 per barrel, market shares collapsed and gold rose to $915 dollars per Troy ounce.

Regulators started looking for scapegoats for the rapid rise. One theory held that malicious traders had spread rumors that BP's diesel refinery in Cushing, Oklahoma was close to empty. If you flew over major cities like Boston, New York and San Diego during those days you saw fleets of oil tankers lying dead in the water. The owners were reputedly waiting for the price of oil to rise before offloading their cargo, so they could make a killing.

By mid-July, prices started to drop as motorists cut back on summer driving. The market was working. Speculators soon realized that the party was over. It was time to buy options to sell oil at $100 per barrel in November.

On September 9, OPEC agreed to cut production to help keep prices at $100 per barrel. That same day, the world's stock markets crashed as brokers realized how much banks and hedge fund managers had speculated on the unsustainable housing bubble.

By early 2009, oil was again selling for about $39 a barrel and Tony Hayward had to inform his board that unless the price of oil rose, all BP's investments in new oil fields would be unprofitable. Saudi Arabia's oil minister, Ali al-Naimi, agreed with the gloomy assessment and pledged that Saudi Arabia would maintain production to keep the price of oil at about $70 per barrel. This would both allow the Middle East oil-producing countries to keep their high incomes and also give oil companies the economic incentives to keep looking for new oil.

Everyone had learned an important lesson, that about $70 a barrel was the optimum price for oil-producing countries to earn a profit and for petroleum companies to invest in further exploration for oil. BP was now in position to continue its hunt for new, more expensive, offshore elephants.

Chapter Four
New Orleans
2008

Since 1945, all the major oil companies had been extracting oil from the Gulf of Mexico in waters less than 1,500-feet deep. In 1985, Shell drilled three successful wells in water twice that depth, proving both that oil could be found there, and that pumping it to shore in pipelines laid along the sea floor could yield a handsome profit.

In 1987, Shell shifted its focus to the Mars site, a much deeper field, 130 miles southeast of New Orleans. The price of oil had slumped again and Shell was short on cash, so it had taken up Jack Golden's offer to have BP pay for 66 percent of the exploratory costs, in exchange for 30 percent of any future profit from the venture.

Shell struck oil in the Mars field, which, ironically, saved its old rival from bankruptcy in 1991. Shell's Technology Manager, Dean Malouta, regretted the decision, "We were crazy to give BP a lifebelt. They brought nothing to the table except money."

What BP had wanted to bring to the table was new technology. After the Mars discovery, the search for deep sea oil had come to a standstill. It turned out that Shell's first successes had been little more than lucky accidents.

The problem was salt.

The Louann Foundation causes sound waves to ricochet back and forth between formations of sandstone and shale, making seismic images of the site look like photographs taken through frosted glass. BP's head of technology, David Jenkins, had promised John Browne, "You'll find more Mars-like oil fields once we can see through the salt."

BP's head of exploration, David Rainey, figured that the way to see through the salt was to stop concentrating so much on geophysics and return to geology. Geophysicists had formerly recorded seismic waves from ships trailing seismic sparkers half a mile apart. But Rainey's team developed software to produce 3-D images from seismic echoes recorded from cables towed only 12 meters apart.

This allowed for much greater resolution, so scientists could see through the salt to the rocks below. This was done in BP's High Visualization Environment, a specially designed, high-tech movie theater in Houston, nicknamed the HIVE.

Rainey was the queen bee of the HIVE. He sat in the front row, center seat of a cinema, surrounded by his workers. Twelve men and one woman would don bulky, battery-operated headsets and stare for hours at images of the earth's plumbing, miles below the water's surface.

These were BP's masters of the underworld — the "Big Brains" Rainey had hand selected for their expertise in geology and geophysics. Rainey would exhort his HIVE mates to think like molecules of oil, slowly rising up through the rocks. They couldn't actually see the oil, but they could visualize the path it took before becoming lodged in a cavity beside a mile-high column of salt or being trapped beneath an impermeable dome.

But was that column really salt or was it sandstone? If it was sandstone, the oil could already have leaked away, along with another $100 million gamble. Too many hundred million dollar mistakes could cost you your place in the HIVE's inner circle.

It was a heady experience for young geologists, many of them just out of grad school.

After sessions in the HIVE, the team would repair to a nearby conference room to argue about where to spend over a billion dollars of BP's money.

When they actually started to drill an exploratory well, the HIVE mates would huddle around monitors in the operations room, watching the computer-guided drill bite down through silt, sandstone and salt.

In the evenings, the onboard geologists would call their HIVE mates at Starbucks to report their findings, which, for a long time, were that the rock samples were still only 3.6 million years old. Everyone knew you had to find rocks over 14.7 million years old before you could feel comfortable that you were in the realm of the sweet Louisiana crude.

At night, the onboard geologists could feel the shake and rattle of the drill as it bit through ever more resilient strata of rock. All their shorebound HIVE mates could do was flip on their laptop computers for one last hit of data before going to bed.

But on the Fourth of July, 1999, all the HIVE mates' work paid off. They had struck a billion-dollar reservoir of oil, the largest ever discovered in the Gulf of Mexico. However, they had to keep their success secret for several months, so BP would have time to buy up all the surrounding sites for a pittance of their worth.

HIVE members were traditionally given the privilege of bestowing secret names on the drill sites so they wouldn't be recognized by competitors, who might overhear them talking while sitting at a nearby restaurant table.

HIVE member Cindy Yeilding was given the privilege of

naming Block 778. She called it Crazy Horse, after Neil Young's band. But demonstrators from the Lakota Sioux tribe claimed BP had denigrated the memory of their chief. The HIVE mates ended up renaming the site Thunder Horse, the same name that caused Browne such embarrassment in 2005.

By March 2008 the HIVE mates had decided BP should buy a lease for Block 252, the Mississippi Canyon in the central Gulf of Mexico, south of Louisiana. The new site was given the code name Macondo, after the fictitious city plagued by a century of rain in Gabriel Garcia Marquez's novel *One Hundred Years of Solitude.*

The auction was held inside the Superdome, the iconic landmark that had witnessed so much tragedy after Katrina and so much excitement when the New Orleans Saints came marching in to their historic Super Bowl win only the month before the auction.

The Superdome sits just up the street from the world-class New Orleans Aquarium. The first thing you see on entering the aquarium is a giant tank containing the three legs of an oil rig. Those legs are surrounded by swarms of ravenous fish. The message is clear; oil, fisheries and tourism are the legs of the symbolic, three-legged stool that supports the economy of southern Louisiana. Who was the lead sponsor of the exhibit? BP, followed by six other major oil companies.

On the day of the auction, the Superdome was teeming with people. There were comely hosts from the federal Minerals Management Service and canny executives from all the oil companies. The atmosphere was reminiscent of the day in 1997 when Chris Oynes, the new head of the Minerals Management Service, opened his briefcase and donned a scarlet jacket. The oil executives had broken into spontaneous applause. They knew the scarlet jacket meant that Oynes intended to continue the New Orleans tradition of "Laissez les bonnes temps roulez!"

George Bush had been urging the Minerals Management Service to "let the good times roll" since his first presidential term. The former Texas oilman knew that the more leases the government sold meant the more Louisianans would get high-paying jobs and vote Republican.

The oil industry had already been credited with saving the Cajun culture. Roustabouts in entry-level manual jobs could make solid, middle-class incomes working two-week on, two-week off schedules, so they could afford to continue their traditional lifestyles of fishing in their justly-famous and prolific coastal bayous.

Chris Oynes had taken the President's message to heart. His employees took hunting trips with oil executives and regularly traded cocaine and sex for industry jobs and private favors.

The corruption ran deep. In 1995, when the price of oil had slumped again, Congress passed a law to expand royalties to oil executives so that they would continue deep-water drilling and production while oil companies' exploratory funds were low.

Yet, Congress neglected to place a price threshold on these royalties. So, when the prices rose to record-setting levels in the mid-2000's, oil companies continued to pay nothing while taxpayers lost out on over $60 billion worth of sorely needed tax revenue.

The Minerals Management Service was also notably riven between its staff geologists, who were always pushing to develop more oil fields, and its environmental scientists, who were always pushing to deliberate until another oil spill contingency plan could be created. Everybody knew which side the head of the agency was on. As if to drive home the point, the MMS gave BP one of its highest safety awards in early 2009.

But the fun went out of the leasing sessions once that old scrooge, Obama, came into office. No more sex and cocaine in the entertainment suites. However, by the end of the auction in 2008, BP had purchased rights to drill wildcat wells in the newly-found Macondo oil field. Executives were ecstatic. Macondo was just the kind of large, gassy oil field they would need to move "beyond petroleum."

BP's environmental impact statement used the standard Minerals Management boiler plate, "Deep sea technology is so advanced it is unlikely that a spill will ever happen, but if it does, the oil can be quickly cleaned up because the coast is more than 50 miles away."

The genie was about to be let out of the bottle, once again.

Chapter Five
"The Kick"
April 20, 2010

Micah Sandell switched off the morning news. The world was still watching that damn Iceland volcano that had hogtied air traffic for almost a week. Reporters were always trying to sensationalize environmental stories.

But Micah had work to do. A helicopter had just flown in a bunch of dignitaries to celebrate Deepwater Horizon's seventh year without a major accident. The Brits were in a jolly mood. BP stock was selling at $59.48 a share and in a few more days they would be moving the rig to a new location.

The only cloud on Deepwater's horizon was that guy, Martin Volek from Halliburton, who kept griping that BP's use of cement was against standard operating procedures, and that a severe gas flow problem could occur if the casings were not cemented more carefully.

Martin Volek didn't have to worry that they were already 43 days behind schedule and it cost $500,000 a day to rent the Deepwater Horizon. BP had already paid Transocean $21 million in extra rental fees.

After he polished off a nice, juicy fat steak in the crew's mess, Sandell climbed into the crane, high above Deepwater Horizon's deck, and switched on his AC unit. It was still warm outside, even though the sun had settled into the placid waters of the Gulf of Mexico almost an hour before.

But the crane operator sensed something wasn't quite right. Perhaps the drill had nicked a deposit of natural gas in the

Macondo oil field two miles below. Perhaps cement from the casing had melted methane hydrates and they had seeped into the drill pipe and started expanding. However it had happened, a massive bubble of methane was now rising rapidly toward the surface, growing bigger and gaining speed as it surged ever closer to the Deepwater Horizon.

At 9:45 a.m., Micah felt a kick and saw a big cloud of gassy smoke erupt from the gooseneck de-gasser that curled downward toward the deck. Seconds later, an explosion ripped through the tank below. Should he climb down into a potential fire? The decision was made for him. A second, larger explosion threw him to the floor.

"No God, no!"

This was what every roustabout feared most; a fire aboard an offshore rig, attached to an infinite reservoir of natural gas. It was like being an ant caught on an outdoor grill.

Somehow, Sandell made it to the lifeboat area. People were yelling, "Drop the boat! Drop the boat!" But not everyone was there, and nobody could do a proper head count because they were so damn scared.

Both Dewey Revette and Micah Burgess had the authority to trip the blowout preventer, which would disconnect the oil rig from the wellhead. Standard operating procedure was to check with Chris Pleasant before activating the system. But the subsea supervisor had already raced to the bridge to tell Captain Kuchta he was going to activate the Emergency Disconnect System.

"Calm down Chris! We are not EDSing."

Thirty seconds later the captain had to exit the bridge and Chris activated the EDS shear rams anyway. That should have driven several hundred pounds of hydraulic pressure against the rams so their massive shears would cut through the drill pipe and choke off the flow of oil. The control panel indicated the shears had worked. In reality they had not.

The tragedy on the Gulf had just begun. The first casualties included 11 of Micah Sandell's rig mates who died in the initial explosion and 17 more who were seriously injured.

The Aftermath

In the ensuing inquiries, all eyes would focus on BP. But of the 126 workers on the rig that day only six employees were from the company formerly known as British Petroleum.

This is certainly not to exonerate BP. It has a long, sordid record of safety and environmental violations. In addition to the 2005 explosion at its Texas City refinery and the 2006 leak at its pipeline in Prudhoe Bay, Alaska, BP had also been accused of secretly dumping waste oil, paint thinners and other hazardous wastes down the annuli, or outer rims, of their drill pipes from 1993 to 1995; and, between 2006 and 2008, three more workers had been killed in three separate accidents at the Texas City refinery.

Despite that dismal record, the spill that occurred on April 20th could just as easily have happened on any of the other 4,000 rigs and 50,000 oil wells operating in the Gulf of Mexico that day.

Deep sea drilling shares many of the risks associated with deep space exploration. It is a complex, high-tech industry that operates under some of the most unforgiving and extreme conditions on Earth. In deep space, you have no light, no gravity and no air pressure. In the deep sea, you have no light, positive buoyancy, 150 times the amount of atmospheric pressure as on land and strong, wayward currents.

I have helped lower a simple oceanographic dredge to the same depths as those encountered by the Deepwater Horizon and watched it return to the surface in an unholy tangle of cable, steel and ruined equipment. All that could be done was to cut the cable and try again. But that work was far simpler than drilling for oil.

The two industries have something else in common. Both the modern era of deep sea drilling and the space shuttle program are about the same age and share the same record of major and minor accidents. In the past 30 years, we have watched the space program lose both the Atlantis and the Columbia space shuttles, and witnessed innumerable close calls with heat shields and ceramic tiles. During the same period, the oil industry suffered the Ixtoc oil spill off the Yucatan Peninsula, the demise of the Deepwater Horizon in the Gulf of Mexico, and innumerable, less-heralded smaller spills and close calls.

But, there is also a major difference between offshore drilling and space exploration. While both ventures are extremely expensive to undertake, there are scant profits to be made flying space shuttles, while the rewards for deep-sea drilling are immense.

There has always been a boom-and-bust mentality to the offshore drilling industry. When the price of oil drops below about $70, adjusted for inflation and at today's prices, companies pull back and stop exploring. When the price of oil surges above about $70, adjusted per barrel, exploration booms. The modern era of offshore drilling started during the 1973 oil embargo when OPEC stopped selling oil to the United States in retaliation for our support of Israel in the Yom Kippur War.

Oil companies suddenly had the cash flow and the incentive to invest in high-tech oceanographic techniques and the new semi-submersible oil rigs that could drill in deeper and deeper water. Later, companies started investing in three-dimensional imaging techniques, like those used in the HIVE, to discover major new, deep oil fields off Norway, Canada, Africa, Brazil, Alaska and the Gulf of Mexico.

The same thing happened when the adjusted price of oil soared above $70, adjusted, during the "second energy crisis," brought on by the Iranian hostage crisis in 1979 and the "2008 energy crisis," triggered by concern about peak oil and the increased use of petroleum by rapidly developing countries like India and China.

All these periods of high-priced oil fueled the present "drill, baby, drill" era of gung-ho offshore oil exploration. Deep-sea drilling has become a game of high-stakes Texas poker with colossal risks and astronomically high potential gains.

During such eras, multinational companies can afford to spend millions of dollars to explore for oil, with the hope that they will make billions of dollars in return. Of course, most of the exploratory wells will come up dry. However, every once in a while, one of the rigs will hit pay dirt. When it does, it means roustabouts can make $60,000 a year straight out of high school, experienced engineers can make $300,000 annually, investors will start receiving nice, plump, little dividend checks in the mail, and billions of dollars will cascade into the waiting coffers of some of the largest vertically-integrated multinationals on the planet. In the words of the shrewd businessman, Warren Buffet, this makes "Mr. Market" very happy.

Of course, it is difficult to generate concern for safety or environmental concerns under such conditions. Companies know that the boom times won't last and they will have to spend big money, fast, to outbid their competitors, explore quickly for new oil fields, and race into production before prices plummet once again.

Oversight tends to be lax. Was anybody really shocked to learn that oil company executives and government officials were offering sex and cocaine to facilitate deals during the Bush administration? After all, places like Alaska, Houston and the Gulf of Mexico were the new boomtowns, and drinking, sex and gambling were the traditional rewards for striking a gusher.

For many months now, BP officials knew they had a gusher. In several days it would become clear to the entire world.

Chapter Six
The Swarm Aboard the Boa Subsea
April 22, 2010

It is no doubt intentional that the pilots' room aboard the Boa Subsea looks like the command deck of the Starship Enterprise. Pilots sit in comfortable chairs, resting their hands on armrests festooned with joysticks, switches and levers. Video monitors and sonar screens display scenes taken by a swarm of small, robotic submarines that were initially designed by the Navy to examine Soviet submarines wrecked on the sea floor.

On April 22, the chief pilot sat forward in his chair, intently studying the brilliant white surface of a blowout preventer as it loomed out of the murky, green waters. He eased back on the right joystick and his remotely-operated vehicle, the Millennium, hovered in front of the blowout preventer like an 11-foot-tall construction worker. Only, this worker was operating below the gas-fueled inferno raging aboard the Deepwater Horizon, a mile overhead.

The pilot recognized the blowout preventer from his training in Oceaneering International's deep-water training tanks in Morgan City, Louisiana. It was the standard model blowout preventer, or BOP, owned by Transocean and built by Cameron International in Beziers, France. The BP guys had used the wrong schematics when testing the preventer only a few days before, but the pilot instantly recognized the inlet tube he now had to activate by hand.

He nudged the left joystick, and Millennium's articulated arm unfolded. He opened and closed its claw-shaped hand several times before clasping it firmly on the well-marked shear ram closer. Another tweak and Millennium's wrist started to rotate like a drill. Hundreds of pounds of hydraulic pressure should have then slammed the shear ram into the drillpipe, but nothing happened.

Perhaps the March accident that had damaged a gasket in the blowout preventer had also allowed all the hydraulic fluids to leak into the ocean. Perhaps the ram was hitting an area where two pieces of drill pipe were coupled together so the metal was twice as thick. This older and less expensive model blowout preventer only had three rams; most of the preventers the other companies used had stacks of seven rams spaced several feet apart so that one of the rams was always over an area without a coupling device. The pilot tried using Millennium's own high-pressure intervention package, but again, nothing happened.

During the following days, up to 14 pilots would squeeze into the BOA Subsea's ROV room to control the swarm of 14 separate, remotely operated vehicles. They also had to keep in touch with pilots on two other surface vessels operating four more ROVs, so that none of the many control cables would become entangled. The ROV's usually worked in teams of two or three, but sometimes the entire fleet was involved.

Each submarine acted as a separate construction worker, politely waiting its turn to perform its one, specific task, whether it be tightening a bolt, cutting away wires, shearing a drillpipe or filming the flow of oil. Only, these tasks were carried out in darkness, under 5,000 pounds of pressure, a mile below the ocean surface.

Despite the pilots' best efforts, just after midnight the flaming hulk of the Deepwater Horizon pitched over and swirled a mile down through the water column to the ocean floor. As dawn approached, BP sent out new submersibles to map the sea bottom and send back damage assessments.

Their reports were not good. They had seen large amounts of oil leaking out of the riser pipe that normally carries oil from the blowout preventer to the surface. The next day, Mary Landrieu, the Coast Guard's newly appointed on-scene coordinator, told a reporter for CBS news that, "There is no crude oil emanating from the wellhead on the surface, er, I mean, on the ocean floor." It was emblematic of the confusion reigning at the time.

Overnight, the story switched from the right stuff to the wrong stuff. The public was not aware of just how good the oil industry had become at discovering and producing offshore oil. The oil companies had been remiss about telling this, the good part, of their story. Now the world was about to find out just how woefully ill-prepared the companies are to deal with an offshore spill.

The technologies for cleaning up oil are laughably out of date. In the wake of the spill, all BP could do was start to drill two new relief wells that would take more than four months to complete. They burned the oil on the surface, captured it in ineffective oil booms, even tried soaking it up in nylon stockings filled with human hair, and when all else failed, they used toxic dispersants to sweep the oil under the rug.

BP did have one clumsy high-tech trick up its sleeve, however: a 125-ton, half-built containment vessel that could be used to funnel oil to the surface for disposal. The dome was transported to Port Fourchon for modifications, but the main modification seemed to be to paint the dome white so it would look good going over the side of the surface vessel in the bright lights at night.

But the day after it was deployed, engineers waylaid Richard Lynch in BP's Houston crisis center. "We've lost the cofferdam!"

Chapter 6   The Swarm

"What the hell do you mean, 'we've lost the cofferdam'? Don't give me that! "

Frozen deposits of slushy methane had built up inside the dome, so the heavy cofferdam was rising like a balloon toward the surface vessel. It could have caused another explosion. The maneuver was quickly abandoned. The public never knew how close BP had come to another fatal disaster.

Two days later, BP announced it would try to use a smaller containment cap called a "top hat" to corral the oil, and if that strategy didn't work they would try to shoot a bunch of shredded old tires down the well, an approach appropriately called a "junk shot." The wrong stuff continued when one of the ROV's blundered into the riser pipe and was almost lost.

To compound their problems, executives from BP, Transocean and Halliburton looked like schoolboys pointing fingers after a schoolyard brawl when they appeared before the United States Congress on May 11. Tony Hayward added further insult to injury when he said that the spill was really pretty tiny when compared to the size of the ocean. To American ears that sounded like saying 9/11 was a small explosion that had only affected a few blocks of a single city on the East Coast.

A week later, Chris Oynes hurriedly retired from the Mineral Management Service. Was that old scrooge Obama at it once again, or was it a tacit admission that something had been systematically wrong with the Service for a long time?

Rush Limbaugh weighed in with the helpful suggestion that the spill must have been caused by environmentalists, because it occurred the day before Earth Day and only weeks before Congress was scheduled to vote on Obama's 'socialist carbon tax.' "What better way to head off more oil drilling and stop nuclear plants than by blowing up a rig? … I'm just noting the timing here."

From the beginning, nothing really did quite seem to add up. Two days after the rig collapsed, BP announced that a thousand barrels of oil per day were leaking from the well. However, even the most casual observer could see that there was already too much oil on the surface for that to be possible.

Within a week's time, sheens of oil covered 580 square miles, yet BP was still claiming that only a thousand barrels per day were coming out of the wellhead.

Edward Markey, the Democratic Congressman from Massachusetts, was in touch with Chris Reddy from the Woods Hole Oceanographic Institution on Cape Cod. The chemical oceanographer insisted that the amount had to be at least five times that much.

Markey was quickly becoming the scourge of both BP and the government. He was chairman of the Select Committee on Energy Independence and Global Warming and used that bully pulpit to jawbone BP into releasing the first videos of oil gushing out of the blowout preventer. Just by looking at footage, experts could see that the flow of oil had to be closer to 60,000 barrels a day. That would mean the well was leaking the equivalent of the Exxon Valdez every four days!

A month later, Markey ordered BP to release the daily live feeds of the oil so the public could see for themselves just how much oil was gushing out of the well. But the public never saw most of the oil because the vast majority of it never reached the surface.

Days after the explosion, and even before the oil had reached the surface, the Coast Guard and the Environmental Protection Agency had given BP permission to release dispersants on the sea floor so the oil would stay safely out of public view. The usual practice was to spray the dispersants from planes when the oil was already visible and easier to deal with.

Now, instead of hundreds of square miles of oil that could be burned off or collected on the surface there would be miles-long plumes of toxic chemicals lingering in the water column for months, perhaps years to come.

There was one silver lining. Shares of the dispersant maker Nalco rose 18 percent the day the day the good news was announced!

Chapter Seven
The Lower Marine Riser
Containment Caper
June 1, 2010, 1:00 a.m.

"Damn!" muttered the ROV operator under his breath. The diamond blade that had been neatly severing the riser had just bound up under the weight of the heavy pipe. He was like a lumberjack whose saw has just seized beneath a towering tree.

He tried the old logger trick of changing the angle of the pipe, but nothing happened. The only thing to do was leave the blade in the riser pipe, bring the ROV back to the surface and put in a new blade. But even that didn't work.

Another operator was finally able to cut away the riser pipe with a giant set of remotely-operated shears, but they left a jagged scar. Now, it would be difficult to fit the new containment vessel snugly over the lower marine riser package to get a tight seal.

None of the ROV operators thought the complex, new lower marine riser package containment scheme would really work, but at least it sounded more professional than "top kill" or "junk shot," the schemes to pump drilling muds and shredded tires down the pipe to stop the flow of oil. Tony Hayward had amped up expectations for the earlier "top kill," operation, calling it BP's, "best hope for capping the well." He had even flown over the surface vessels in a helicopter and declared, "We have wrestled the beast to the ground, but we still haven't put a bullet in its head." Perhaps it was supposed to resonate well in American ears, but it just didn't sound right, delivered in an ever-so-fey British accent.

One of the technicians who worked on both "junk shot" and "top kill" admitted anonymously that neither procedure had even come close to succeeding. The pressure of the outpouring oil and gas had been simply too great to overcome with shredded tires and drilling muds. BP had closed down "top kill" after only three days and then had to come up with something quickly to look like it was still in control of the rapidly deteriorating situation. But the "lower marine riser package containment operation," still seemed like a desperate move with a fancy title.

President Obama joined the chorus of criticisms when he was asked on the Today Show if he wished he could fire Tony Hayward. The President responded, "I don't sit around just talking to experts because this is a college seminar. We talk to these folks because they potentially have the best answers, so I know whose ass to kick." The Brits weren't used to having such words fired at the boss of England's largest corporation, one that had given the royal exchequer more that $14.4 billion in taxes in 2008.

But the Obama administration had some much harsher measures in store for BP. On June 16, Obama jawboned Tony Hayward into agreeing to put up $20 billion toward a relief fund to compensate fishermen and businesses for any lost income incurred because of the spill. Two days later, BP's board announced that Bob Dudley would take over day-to-day operations of the cleanup effort from Tony Hayward. It made sense; Bob Dudley had grown up in Mississippi and could put on a deep southern drawl when the situation called for it.

The Obama administration had another trick up its sleeve. It declared a six-month moratorium on deep water drilling. If the moratorium was extended, it could mean that BP would not have enough income to stay in business, let alone to continue paying for the relief fund.

Since the explosion, the Department of Interior's Minerals Management Service had approved 27 new leases for offshore drilling that had already been in the pipeline. They had all received the same prior exemptions from environmental review that BP had received for the Macondo field. Now, they were going ahead despite the moratorium. Kieren Suckling from the British Centre for Biological Diversity said, "This oil spill has had absolutely no effect on the Minerals Management Service's behavior at all. It's still business as usual, which means rubber stamping oil-drilling permits, with no environmental review."

BP was the proud owner of 13 of the offshore leases. It was a testament to their belief that offshore oil was the wave of the future and that the whole moratorium thing would blow over once the well was safely capped and demand for oil reasserted itself. It was a pretty safe bet.

Meanwhile, patches of oil and tar balls were starting to wash up on Florida's shores, in addition to those of Louisiana, Texas, Mississippi and Alabama. BP installed a three-ram, capping stack on the lower marine riser package, which started cutting off the amount of oil flowing into the Gulf. This was the most critical phase of the operation. If the pressure rose too quickly, it could overcome the casing and blow out the entire well area, a situation oilmen called "cratering."

For days, Stephen Chu, President Obama's Nobel Prize-winning science advisor sat, surrounded by a gaggle of scientists, peering at pressure data as BP ran tests on the well. At one point, National Incident Commander Thad Allen asked BP's Bob Dudley to provide written procedures for reopening the well, because tests seemed to show oil seeping some distance from the hole. This could mean the well casing was about to fail and the surrounding sea floor could crater, creating a potentially unstoppable situation. But, to everyone's surprise, the casing continued to hold and BP proceeded to pump drilling mud down through the new cap to try to bullhead the well.

Shortly after midnight on August 4, 2010, BP reported the well had achieved "static condition": drilling mud had filled the entire well. A few days later, BP's board announced that starting October 1, 2010, Bob Dudley would replace Tony Hayward as CEO of BP. He was the same Bob Dudley who had been thrown out of Russia in his early days at BP, but he was an American who had grown up in Mississippi and acquitted himself well throughout the capping process. The immediate crisis appeared to be over.

Chapter Eight
Ixtoc and West Falmouth;
Two Spills, Two Outcomes.
1969 and 1979

When I first heard about the explosion aboard the Deepwater Horizon I was busy writing about the 2010 Icelandic volcano eruption. I thought it was going to be the defining environmental event of the year. Boy was I wrong!

But the oil disaster did make me pause long enough to remember other spills I had witnessed in the past. The first was a small spill in West Falmouth, Massachusetts. It occurred on a foggy September night in 1969 when the oil barge, Florida, broke free from its tugboat line and ran aground, just before entering the Cape Cod Canal.

This was during the first wave of modern environmentalism, when scientists were first becoming concerned about oil pollution caused by oil tankers cleaning their tanks at sea, and ships that didn't have double hulls to prevent spills such as this one.

For the sake of our knowledge about the effects of oil, the spill fortuitously occurred in the hometown of the Woods Hole Oceanographic Institution, affectionately known as WHOI, pronounced as "Hooey" by those in the know. When the spill occurred, the oil industry assured the public that most of the oil would evaporate quickly and that the effects would soon be long gone and forgotten. Their statements reflected the conventional wisdom of the time. But to a group of old-fashioned WHOI scientists, that kind of untested assumption presented a challenge and a perfect case study to test such an unverified hypothesis.

WHOI had a small fund set aside for just this sort of situation. It allowed the scientists to start collecting samples from the first day the diesel fuel began washing ashore. It also provided them with an economic bridge to tide them over until they could apply to the National Science Foundation for further funding.

WHOI scientists George Hampson described his experience from the first day of the incident. "The odor of Number 2 fuel was all over North and West Falmouth. Dead fish and lobsters were washing up everywhere. Three-foot-long lugworms that normally spend their lives buried deep under the sand were poking through the surface of the oil-slicked muck. Thousands of shellfish emerged from the ground with their necks extended as if gasping for breath. Normally, you would hear sea gulls and crickets at this time of year, but there was nothing. We started calling it, 'The Silent Autumn.'"

The team didn't just collect the large, obvious animals like birds and fish; they used bottom grabs and plankton nets to meticulously collect every tiny, dead animal that lived in the water column and on the seafloor. They even collected the microscopic, interstitial animals that lived between the sand grains of the beaches and marshes.

Despite intense efforts by the oil industry to stop their work, the scientists persevered, continuing to collect samples for several decades so the heft and integrity of their data grew ever more robust. When a senior scientist retired, his graduate student, and ultimately, even his great, great, grand-graduate students would continue the research. Eventually, scientific papers about the West Falmouth oil spill became the gold standard for how research on oil spills should be done. They also helped set national standards for how to clean up oil pollution once a spill has occurred.

The WHOI scientists discovered that, immediately after the West Falmouth oil spill occurred, there was a massive die-off of almost every living creature in the area. Even mobile animals like lobsters died because they were attracted to the smell of oil. This came as no surprise to local lobster fishermen who used to put bricks, soaked in kerosene, in their traps to attract their odor-sensitive prey.

A few seasons after the initial die-off, the diversity of animals remained low. For several years, about the only animals that could be found were a few species of polychaete worms. The seafloor was still severely unhealthy and out of balance.

However, their most unexpected discovery was just how long the oil remained in the substrate. Today, 41 years after the spill, you can still drive down to West Falmouth, wiggle a shovel around in the mud, and watch pure, unaltered oil ooze to the surface. The shellfish beds remain closed, and fiddler crabs still dig their burrows at right angles to avoid the oil, lurking only inches below the surface. They also continue to stagger around as if drunk, making it easy for predators to prey on these crabs whose burrows help oxygenate the marsh.

It turns out that the entrance to the Cape Cod Canal has more oil spills than almost any other place in the United States. Tugboats must use the canal to haul fuel barges from New York to Boston, even during New England's occasional hurricanes and perennial winter nor'easters.

Five years after the fuel barge Florida ran aground in West Falmouth, the Bouchard foundered only two miles away. Nearly 30,000 gallons of toxic fuel oil covered the marshes of Winsor Cove, killing its lush beds of *Spartina* saltmarsh cordgrass. "The first year, it was just a moonscape," recalled George Hampson.

The grass, which served to keep the underlying peat beds from eroding, did not recover quickly enough to get a hold on the beds before they crumbled. "It was a race between how much peat from the marsh was eroding and how quickly the grass was coming back," said Hampson.

For several years, the dying marsh continued to erode, and still struggles today against the tides. "It's been 35 years, and I'd say the grasses are just beginning to grow back," concluded Hampson. Will that be the fate of the Gulf of Mexico?

Another spill provides more insight into this question. It is the Ixtoc spill that occurred off of Mexico's Yucatan peninsula in 1979.

I had been invited to film the spill by the publisher of an influential oil spill newsletter. I flew into Mexico with my many naïve assumptions firmly intact. The shrimp fishery had to be a little mom-and-pop operation, where fathers passed their boats on to their sons and everyone made a small, but honest, living. The fishermen were the good guys who wore the white hats and the oilmen were the bad guys with the black hats.

I was quickly disabused of such simple notions when we first drove into the city of Merida. There, in the city center, stood a fountain with a 25-foot-high bronze statue of a shrimp that was being constantly spritzed with lush cascades of bubbling, streaming water. I soon learned that Mexico's shrimping was huge, second only to the oil industry. Its fleets were owned and operated by international companies, many of them headquartered in Norway. The companies would put $400,000 down on a new shrimp boat and have it pay for itself in only six, short months. Obviously, there was a lot more riding on the outcome of this spill than just what was floating on the surface.

I also witnessed the influence of oil money on a small, local, sustainable economy. On my first night in the city I joined a table of American oilmen and sophisticated, young American-trained executives from Pemex, Mexico's nationalized oil industry. One of the oilmen boasted that he had just flown in from Houston and didn't have a clean shirt, so he had given a hundred dollars to a 12-year-old boy to take his dirty shirt to the laundry!

Our Pemex hosts recommended that we just order the shrimp hors d'ouvres, because they would probably be enough for dinner. Sure enough, giant platters arrived, overflowing with big, succulent, steaming shrimp. If I remember correctly, it cost less than five dollars to get more shrimp than you could ever want to eat in one sitting.

The conversation drifted to the Diego Rivera mural of the Los Niños in the Palacio National, in Mexico City. As you enter the room you look up and see a startling mural of six students wrapped in Mexican flags falling headfirst, directly toward you. The Texas oilmen joked about the depiction and I could see more than one of the Mexican executives, who spoke perfect English, flinch, but they all held their tongues.

The mural represents the six student cadets who fought in the 1847 occupation of Mexico City during the Mexican-American War. Though greatly outnumbered, the 15- and 16-year-old boys refused the order of their general to retreat, and bravely fought the American enemy. It is believed that one of the cadets wrapped himself in a Mexican flag before jumping to his death from the ramparts of what is now the Chapultapec Castle. It's no wonder that American oilmen are not always welcomed with open arms.

I also returned from the trip with a bit of an international scoop. New England was in a political battle over whether to allow oil companies to drill for oil on Georges Bank, the world's fourth largest and most productive fishing grounds. As soon as the Ixtoc spill occurred, the companies were quick to assure the public that such an accident could never happen on an exploratory well and in any case it would never, ever happen on an American oil rig.

But, as I was flying over the rig, I happened to film a small decal of an American flag attached to the side of the Ixtoc rig. It wasn't anything official looking, it just seemed like someone had reached out of his porthole and stuck the decal to the side of the platform. But this was curious. It was supposed to be a Pemex rig with an all-Mexican crew.

Later, at the airport, I overheard an American roustabout making plans to stay in his cabin in the Rockies until the spill was over. When I returned home I did some research and discovered that not only was the platform an exploratory rig but it was also partially owned by an American company. I wrote an Op-Ed piece for *The Boston Globe* that caused a bit of a stir, and Senator Kennedy's office invited me to Washington to show the footage at a meeting on the Senate floor.

Later, it was discovered that not only was the Ixtoc rig partially owned by an American company, but that a quarter of the company was owned by the then-sitting Vice President, George Bush, Senior. After the discovery, the rig was hauled out to sea and sunk, rather unceremoniously and a little too quickly for comfort. It had been a fun little scoop, for a still wet-behind-the-ears science writer.

Ixtoc was the worst oil spill in history, and it held that dubious distinction until the BP spill occurred, 31 years later. The sweet Mexican crude gushed out of the shallow Ixtoc well and burned on the surface for over 10 months. Eventually, the colorful oil-fire expert Red Adair flew down to Mexico and capped the spill with a containment dome that became known as the "Mexican Sombrero."

After the oil was contained, people looked around and wondered, "What happened to the 140 million gallons of oil that spilled?" In the following years the shrimp industry boomed, much as it had for generations. The answer seemed to lie in the oil itself. Unlike the lighter diesel fuel that came ashore overnight and insinuated itself between the sand grains in the Falmouth marsh, the Ixtoc oil was a light, sweet crude that evaporated on its own and could be burned at sea. It also weathered into relatively inert tar balls that could be scooped up off the ocean surface. Offshore winds had mostly blown the weathered oil away from the shrimp's fragile nursery grounds in the mangrove swamps, but most importantly, the water was warm and teeming with bacteria, evolved by nature to feed on oil that seeped out of the seafloor along this rich Campeche coast.

So, which oil spill would provide the best answers to understand the spill in the Gulf of Mexico?

Chapter Nine
The Plume
Aboard the R/V Endeavor
June 28, 2010

Rick Camilli lay in his bunk, trying to gauge if the swell had increased overnight. This was his fifth day on station over the plume of hydrocarbons spewing out of the BP well. Now, hurricane Alex was bearing down on the Gulf of Mexico and they might have to quit their research before they had a chance to fully map the plume. Better head up to the bridge and check with Rhett McMunn.

"Good morning, Captain. How's the weather look?

"Not so good, Rick. The Weather Service has Alex blowing into the Gulf in two days as a potential Cat 3 hurricane. Coast Guard wants us out by tomorrow."

"So, do you think we can squeeze in one more day of research before heading back to Florida?"

"As long as you can get your bug hunters out of the sack before breakfast!"

The chief scientist smiled. The crew always referred to all scientists as bug hunters, even though on this cruise, most of them were chemical oceanographers.

Chris Reddy was satisfied with the news. Everyone was dead tired and ripe. They had only had time for catnaps, and nobody had taken a shower for the past week because their water desalinizer had been plugged with spilled oil.

The co-principal investigator could have used several more days to fully characterize the plume. He knew it was chock full of hydrocarbons, and was at least 22 miles long. But, it would be nice to confirm the actual length. This sure wasn't like the early days when he cut his teeth on using gas spectography to characterize the West Falmouth spill. Back then, all you had to do was drive to the site to get your samples.

Rick and Chris were amazed at how quickly they had become involved with this crisis. On May 19th, Rick got the call from the Coast Guard. He had been about to step on a plane headed to Australia, where he was supposed to captain a boat during a long-planned family vacation on the Great Barrier Reef.

The Coast Guard was in charge of the operation, since the Deepwater Horizon was technically classified as a ship because it was a free floating semi-submersible rig. The Coast Guard expected a lot of oil to surge out of the well, and asked if Rick could help them calculate the amount. Just 24 hours later, he found himself in Louisiana, feverishly trying to calculate flow rates instead of sailing leisurely over the Great Barrier Reef.

The fumes were so awful Rick had to lift his respirator out of the way so he could see the pictures his wife sent him from Australia. Some birthday! But, his flow estimates had made him household name. Overnight, he had gone from being a quiet, Woods Hole researcher to a rock-star scientist, who was being asked to appear before Congress and getting daily requests for news interviews.

Rick and Chris had pulled this cruise together only a week after hearing they had won a RAPID grant from the National Science Foundation. RAPID grants were designed to put scientists in the field with a minimum of bureaucratic hassle. Some wonderful soul at the National Science Foundation had even prevented bureaucrats from coming up with any jargon-filled names to fit the acronym. The grants were simply flagged

"RAPID" to get them out the door to address situations as quickly as possible.

As soon as the WHOI administrators knew the grant was in the pipeline, they advanced Rick and Chris money so the two could find a research vessel capable of handling their specialized gear.

All the WHOI vessels were already committed to other cruise tracks , but the University of Rhode Island agreed to cancel the research ship Endeavor's trans-Atlantic mission and redeploy it into the Gulf of Mexico.

The only problem was that the vessel was too cramped to provide bunks for research assistants or graduate students. This would be a cruise with five principal investigators. It was like a movie with five stars and no supporting actors.

Rick ripped the hurricane report out of the fax machine and strode to the main deck. The day was hot and humid, and long, greasy swells were already rolling out of the sultry southeast.

Rick signaled to the winch operator, and the five senior scientists pushed and pulled the remotely-operated underwater vehicle, Sentry, over the side. In 24 hours, the waves would grow large enough to tip the crane in and out of the water, making it too dangerous to retrieve the vehicle, which was roughly the size of a bus. One sweating scientist could be heard, sotto voce, "Where is a graduate student when you really need one."

Unlike the ROVs the oilmen used, the Sentry was an autonomous underwater vehicle that could be pre-programmed to cruise back and forth through the plume on it's own, and it carried power for several weeks of operation.

The Sentry would help the scientists decide where to lower the large, round arrays of water-sampling bottles and

Thethys, a shoebox-sized mass spectrometer that measured concentrations of oil in the water. During the West Falmouth spill, Chris had to drive the water samples to WHOI's huge mass spectrometry lab. Here, Tethys could analyze the plume and send the data back to the Endeavor in real time.

Scientists from other institutions had tried to detect and measure the plume, but they had been foiled because their equipment could only measure the water column vertically. But Sentry could scan the water column horizontally to provide precise coordinates, so the researchers could target their vertical profiling equipment with pinpoint accuracy.

The scientists squeezed into the lab and huddled around the computer monitors. Tiny blips on the Tethys screen showed discernable amounts of hydrocarbons, which contained benzenes, toluene, ethylbenzene and total xylenes in excess of 50 micrograms per liter.

You couldn't actually see the hydrocarbons in the water. Chris Reddy said, "It wasn't like a plume of Hershey's syrup down there." But you could see a definite signal of hydrocarbons and, perhaps, dispersants, leaking from the wellhead area. The plume was 1.6-miles wide, 400-feet high and hanging about 2,000 feet below the surface.

One of the scientists explained that large droplets of oil have enough natural buoyancy to rise to the surface, where they can either weather into tar balls or dissipate into the atmosphere. The problem with dispersants is that they break up the oil into tiny droplets so small that they lose their natural buoyancy. So, most of the hydrocarbons in this plume would remain in the water column for at least a year. This meant that invisible pulses of hydrocarbons would continue to come ashore for months, if not years.

The scientists also discovered that there weren't any dead

zones around the plume. Dead zones indicate that bacteria are chewing up the hydrocarbons and consuming oxygen. So, while it seemed like a good thing that there were no dead zones, in reality, that meant it would take bacteria much longer to decompose the hydrocarbons.

The scientists had encountered an initial problem with the oxygen samples. As oil collected on the sampling equipment, the equipment started to give false readings. The scientists solved the problem by switching to the old Winkler Titration technique. This was oceanography the way it had been done 50 years before, giving the largest number of readings and the most irrefutable evidence.

When the WHOI paper was published in August, BP officials tried to dispute its findings by arguing that the oil was just naturally making it's own, slow way toward the surface.

It looked like the West Falmouth spill all over again. WHOI scientists had carefully and clearly measured a 22-mile-long plume of hydrocarbons in June that was probably still there in August.

This was a highly-inconvenient truth. Both BP and the government wanted to be able to declare the emergency over. The National Oceanographic and Atmospheric Administration monitored the area and stated that they had failed to detect any "actionable" concentrations of oil.

Their failure to find oil was far more convenient. The oil-spill response team was in the process of laying off 23,000 clean-up workers, and Paul Zukunft was replacing Thad Allen as point man on the project.

The final results were somewhat ambiguous, because nobody had the money to go back and sample the site with the same equipment. Unlike in West Falmouth, where scientists could simply drive back down to the site and get more samples to bolster their argument, there was no money left to return to the gulf with Sentry and Tethys and see if the plume still lurked 2000 feet below the surface.

Chapter Ten
The Flock at Crane's Beach
Ipswich, Massachusetts

It is September 30, 2010. Layers of gray-and-white clouds scud across the evening sky. Before doing yoga, I lie back in the sand and become lost in a hypnogogic kaleidoscope of criss-crossing clouds. All sense of stability disappears. I am floating through a misty miasma of vaporous clouds.

Suddenly, the dark image of a plane emerges from the clouds, circles wide, then plunges back into the void for its final descent toward Logan airport, 30 miles away. The airlines are rushing to land as many planes as possible before the East Coast gets socked in by the remnants of hurricane Nicole.

I am not alone on the beach. Flocks of seagulls mill about, just overhead. But something isn't quite right. The seagulls seem to fly along naturally enough, and then they stall and stretch their bills into the sky as if they are about to crash. After a while, I realize that the gulls are actually snatching tiny, insects out of the evening air. Judging from their boisterous calls, they are enjoying themselves immensely.

The insects seem like a welcome addition to their usual diet of moon snails, surf clams and stolen sandwiches. The gulls are so intent on capturing the tiny delicacies that they don't even notice me. One of them almost flies into my face, and I'm tempted to pluck another out of the sky as I reach up to do my sun salutes. It would not be very yogic, but perhaps just a little bit zen.

A mixed flock of plovers and sanderlings sit on the far end of the long, thin sandspit where I am doing my poses. Every once in a while, the flock is seized with a fit of nervousness, and the birds wheel out over the ocean, flashing gray, then brilliant white as they dash back and forth like a school of frightened minnows. It is enough to wake me from my quiet trance. I wonder why they expend so much energy on such a brilliant display— certainly not just to delight a human doing yoga in the surf.

The flock arrived early this morning, after having flown all night from its nesting grounds above the Arctic Circle. They spent their first few hours on the beach, their tiny legs a blur as they dashed in and out of the waves, probing ravenously for interstitial animals. As the tide rose, the birds retreated to the upper beach where they slept all day to prepare for the next overnight leg of their 4,000-mile flight to the shores of Central and South America.

A few minutes of observation reveals the cause of the flock's unease. A Peregrine Falcon is terrorizing the shore. His swept-back wings propel him up and down the mile-long strip of dunes, beach and grass. He rises as he approaches the sandspit, focuses on a single bird, then dives toward the hapless animal at 200 miles per hour.

The flock flashes brilliant white as the birds dash to the left, then cut back to the right with only their grey backs exposed. The ruse works. The Peregrine loses focus on the single bird and rises to try again. This time, it flushes a single bird from the flock, and I expect it will come down with its prey, but somehow the bird manages to fly over the falcon and escape. To tell the truth, the young falcon is not very good at this style of hunting. It seems to be confused and irritated by the Sanderlings tactics. Now their display makes sense. Of course, it is not just for my delight, but a brilliant strategy to make the falcon lose its focus.

As I head back down the beach, I see a display that is,

perhaps, just a little bit more for my benefit. A striking young woman with masses of wild black dreadlocks rides bareback down the beach on a 15-hand-high, galloping charger. I know the rider. She is an emergency room physician at a nearby hospital, but from now on I will think of her as the warrior goddess she really is.

Suddenly, a siren wails out over the dunes. The Crane's Beach rangers are signaling that it is sunset and they are about to lock the parking lot for the night. As soon as they hit the siren, 20 coyotes start howling from the top of the dunes, only 20 feet away. I decide I would rather deal with a group of disgruntled rangers than a troop of erratic coyotes packing miscellaneous red wolf genes.

I hustle down the wrackline and see the Peregrine Falcon one last time. He is ripping feathers off the breast of a decapitated young Piping Plover. Apparently, he figured out how to hunt shorebirds after all.

Behind me, the flock is again beset with the same unified agitation. Several times, the birds wheel out over the water and back again. Perhaps they are testing the wind direction to see how well they will be able to fly in this incoming cold front; perhaps they are reorienting themselves to the earth's magnetic field that will guide their flight southwards. Perhaps another falcon is looming near.

Eventually, the entire flock flies out over the ocean, circles twice and proceeds purposefully south. They will fly all night through the remnants of hurricane Nicole, which has already killed nine people in Mexico.

A few days later, the flock arrives on the shores of Padre Island, Texas, but something is wrong. There are no interstitial animals and the sand is thick with oil. The entire Gulf of Mexico is covered with a sickly, iridescent sheen of greenish dispersants. Millions of gallons of dispersants have bathed this beach for several months. They have filtered down through the sand, killing wantonly and leaving a deadly residue in their wake.

Some of the birds will sicken and die, others may fly inland, where the state of Louisiana has paid crawfishermen and rice farmers to flood their fields and ponds to provide clean wetlands for the famished birds. These will be the fortunate few that will be able to continue their southward migration to the beaches of Chile.

Chapter Eleven
The Peninsula
November 9, 2010

The dew was still heavy on my windshield as I pulled out of the Riverside Hotel parking lot in Belle Chasse, Louisiana. As I drove south on Interstate 23, the sun shone down on the 80-mile-long peninsula that stretches like a snake's tail from Belle Chasse, on the southern edge of New Orleans, to the tip of the Mississippi River Delta.

The Mississippi flowed behind tall levees to my left, and the Gulf of Mexico sat behind high levees to my right. Orange groves and cattle ranches flourished between the levees on rich, organic soil made from the accumulation of over 12,000 years of sand and mud from the Mississippi River mixed with the accumulation of over 12,000 years of limestone shells from the Gulf of Mexico.

I passed through towns with evocative names like Jesuit Bend, Potash, Port Sulphur, and Pilot Town. A small helicopter company shuttled roustabouts back and forth to offshore rigs, while massive piles of grain and coal, just off-loaded from cargo ships, lay beside the railroad siding, waiting to be loaded onto freight cars for distribution throughout the country.

There were also reminders that this place has produced oil and fish for over 75 years. An old-fashioned, nodding-donkey oil derrick sucked oil out of the ground with every dip of its metal head. Pipelines emerged from the marsh carrying offshore oil to a Chevron refinery that manufactures the fuel additive Oxonite. I watched as they pumped the valuable liquid into long black tanker cars that rumbled past my hotel window every afternoon at 4:00 p.m., sharp.

Beside the refineries, thousands of shrimp and fishing boats streamed out of the peninsula's bayous for another day of fishing in the productive waters of the Gulf of Mexico. Only the oyster boats lay idle. Most of the oyster beds had been killed off when fresh water was diverted into the marshes to help push the oil back out into the Gulf of Mexico.

These human activities were but a reminder that despite hurricanes and oil spills, this peninsula is a resilient place. Its economy rests on food and fuel, the two basic commodities that make modern civilization possible. As long as you know how to set a net or stack a drillpipe, you can make a good day's pay and enjoy an enviable lifestyle in the bustling communities of this peninsula. Unlike areas dominated by manufacturing, such as Detroit or the Rust Belt states of the Ohio Valley, this place should remain a haven for blue-collar jobs for years to come.

I had timed my trip to coincide with the six-month anniversary of the spill. Scientists needed at least that much time to fully understand the subtler ramifications of this complicated tragedy. There were indications that the oil had insinuated itself into the food chain and that we would probably not see all the effects of it until the next generation of fish, crabs shrimp and oysters hatched out in the spring, if not four or five years later.

There were other indications that the oil was killing species of deepwater coral and causing dolphins to abort their fetuses. The biggest question was whether the heavy use of dispersants had hurt or helped the cleanup effort. Dispersants sprayed from planes had undoubtedly kept much of the oil out of the marsh, where it could do the most long-term damage, but how much had the underwater spraying of dispersants affected plankton at the bottom of the food chain? Had it been simply a ploy to hide the spilled oil from the public, or a clean-up technique that had worked better than was anticipated? This was one of many subtle questions that I hoped to resolve while down here, if it was yet possible to do so.

I had arranged to join a Plaquemines Parish clean-up crew still working to suck oil out of the marsh in Bay Jimmy. But the day before we were to leave, I checked with my Plaquemines Parish contact and was told my trip had been cancelled. BP had closed down the clean-up effort in Bay Jimmy and moved the headquarters of the joint operation south to Venice, Louisiana, closer to the tip of the Mississippi Delta.

This was a kick in the pants. I had organized my entire trip around seeing the clean-up operation, so I decided to drive down to the Bay Jimmy staging area anyway. Perhaps, I would be able to hitch a ride to the site on a private boat.

The first person I saw was a blue crab fisherman stacking his traps. He was putting them into storage because BP was still paying him good money to finish cleaning up after the spill. He recommended that I just continue driving down to Venice and try my luck there. Even on the highway, evidence of the clean-up effort was disappearing fast. Workers were already tearing down billboards that lawyers had put up to advertise their legal services to those who had lost income during the spill.

When I arrived in Venice it was clear that BP was running their Vessel of Opportunity program with military-like security, if not military-like efficiency. They had commandeered all the available boats, and you could only get permission to see the cleanup if you made a request several weeks in advance. I recognized a dodge when I saw one.

The trip had allowed me to see this valuable peninsula, but there was still a major fly in the ointment. For years, scientists have known that marsh depletion was one of the most serious environmental concerns facing Southern Louisiana. The problem is that the Mississippi River wants to writhe back and forth like a squirting garden hose. Today, the Mississippi is about as far east as it can go. It would like to writhe back west, spewing its valuable sediments onto another section of this coast as it does so. But instead, it is straight-jacketed between two rigid, manmade levees that force it to shoot that valuable load of sediment far out into the Gulf of Mexico where it settles beneath the deep waters of the Mississippi Canyon, and can do no good.

The Mississippi River would also like to hook up with the much younger Atchafalaya River and flow south into the Gulf of Mexico, west of New Orleans. That would rapidly rebuild the marshes of the Atchafalaya Basin. The problem is that while joining the two rivers would build up the marsh and help reduce the impact from future hurricanes, it would also leave the city without the Mississippi River, and all the towns and facilities along this peninsula would have to be moved to a new Mississippi Delta—not a very likely scenario. That is why the Army Corps of Engineers spends millions of dollars a year to prevent the confluence of these two rivers, as Nature would have them do.

Another intriguing, new potential solution has emerged from this oil spill. Halfway through the catastrophe, Governor Jindal convinced BP to give the State of Louisiana $360 million so it could construct sand berms to prevent the oil from reaching the marshes. Now, through the vagaries of funding and politics, this hare-brained idea was being transformed into a potentially state-of-the-art project to restore Louisiana's invaluable marshes. I would see the problems and their possible solution the following day in the marshes off Delacroix, Louisiana.

Chapter 12

BP's Surprising Legacy
Delacroix, Louisiana
November 10, 2010

"Hold your horses!" Michael smiled at a skein of ducks a Northern Harrier had just scared up off the marsh. "I have a little date with them next Saturday when the season opens," he explained, as he throttled down our ear-deafening airboat. "I can already taste 'em in my brother's duck-and-sausage gumbo."

Even if the government didn't pay him, there was no place Michael Bell would rather be than on this marsh. Ever since he was a boy, Michael had been fascinated with marshes. He loved the smell of the mud and the way it felt when it oozed up between your toes, and through your fingertips when you squeezed it in your hands.

"I spent most of my summers on the water. My parents were pretty lenient—only firm rule we ever had was: 'no kids allowed in the house during the day.' Why waste money on air conditioning when you could be outside? That suited me and my brother Brad just fine. We would be up every morning at dawn and out in a boat by daylight.

"It really didn't matter what kind of boat it was. I was the oldest, so my brother had to paddle, anyway. We'd just see a nice pirogue and say, 'Hmm, looks like a pretty good boat: nobody's using it, so we might as well take it out fishing.' We would spend all day building forts, blazing trails and catching bass, brim and milk-mouthed white crappies, only we called 'em 'sac-a-lait.

"We started to fish from motorboats after Dad bought a store on Cypermort Point, near New Iberia. I guess you could say we were just coonass Cajuns, living the life of Huck Finn and Tom Sawyer; fishin', froggin' and catchin' alligators. We used to use headlights to jig for frogs. The backs of their retina would reflect the light, so you'd look for these glowing eyes staring up at you. You'd learn pretty quick to only spear the green eyes; the red eyes belonged to alligators, and you didn't want to have to wrestle one of those in the dark.

"I was always just on the edge of getting into serious trouble as a kid. It was usually for speaking French in class. My mom and dad both spoke French at home, so I was always getting sent to the principal for talking French behind the teacher's back.

"After high school, I joined the military and became a hospital corpsman. After leaving the military, I took pre-med classes at the University of New Orleans. That worked out just fine until hurricane Katrina came along and destroyed my house in Lake View. That's when I decided to return to Baton Rouge and try to get back into medical school.

"One day, after classes, I went to talk to my organic chemistry professor about stereoisomerism, or some such mind-boggling concept. Fortunately, he didn't have much to do that afternoon, so we had a good long talk. Finally, he turned to me and said, 'You know, Michael, we don't need another doctor sitting inside a stuffy old hospital working with sick people. We've just been through the worst hurricane in our history. Louisiana needs you to be outside, doing what you know best.'

"He suggested I walk over to the School of Renewable Natural Resources and see what they had to offer. I really owe Dr. Crowe. Studying the science of nature had never occurred to me. That afternoon, my major changed. My first job out of LSU was with the Department of Interior. That's what got me out on this boat, collecting marsh data."

What Michael and I couldn't see that day were any signs of the oil spill, just hundreds of ducks, geese, fish, and thousands of acres of marsh grass waving in the quiet breeze. What we could see, however, was how much this marsh had been degraded by over 75 years of abuse.

Unlike other coastal states, Louisiana has no beaches to speak of. It is flanked, instead, by about a hundred miles of marshlands that stretch from the mainland into the Gulf of Mexico. From the mainland side, the first 50 miles of marsh are flooded every spring by fresh water, and consist of freshwater plants. The second 50 miles of marshland get flooded by salt water on every tide, and consist of saltwater plants. A northerner can see the difference right away. Alligators leer up at you from creeks in the freshwater marsh, but the saltwater marsh looks just like those you might see on Cape Cod or Long Island Sound.

As soon as we passed Mozambique Point, Michael pointed to the trunks of dead oak trees jutting out of the upland marsh, "Those have been killed by saltwater intrusion."

He explained that thousands of miles of channels for oil pipelines crisscross through this marsh like spaghetti, carrying oil from the offshore oil rigs to the mainland. When the oil companies originally dredged these channels, they were only about 30-feet wide. But the channels allowed salt water to flow into the brackish water interior of the marsh, where it killed the fresh water plants and started to erode the edges of the channels. Today, most of the channels are hundreds of feet wide and only a few, small, frayed patches of marsh remain in a shallow sea of mud.

The other problem is that the soil under the marsh is compacting because it is no longer being replaced by sediments from the Mississippi River. The river now flows through the straitjacket of artificial levees that shoot the valuable sediments out into the deep waters of the Gulf of Mexico, where they fall, uselessly, to the ocean floor. These combined problems are causing this coast to lose the equivalent of a football field worth of land every half hour. By the time we return to the dock, some of the houses along this bayou may have lost another foot of lawn off their back yard lots.

The first data-sampling station revealed the problem. We had traveled about 40 miles from the freshwater marshes near Delacroix, towards the edge of the Gulf of Mexico. You could see the silhouettes of oil rigs looming in the distance. We pulled up to a simple, wooden-plank boardwalk that jutted into the marsh. It was covered with blue crab and oyster shells and scat from a raccoon that sat on the boardwalk the previous night, enjoying its own tasty seafood gumbo.

The boardwalk reached into thousands of acres of waving *Spartina* saltmarsh cordgrass, flecked with the white-and-purple blossoms of late-blooming marsh asters. *Spartina alterniflora* is one of the only grasses that can tolerate the high salinity of this water, which is about half that of the open ocean only half a mile away.

The second station, located several miles inland, was more interesting. The salinity was about a third that of the open ocean and the vegetation alternated between saltwater *Spartina* grass and freshwater wiregrass. Michael explained that when fresh water floods through the marsh it allows the wiregrass to flourish, but when the rain stops it allows the saltwater tides to kill off the wiregrass and the *Spartina* grass to reestablish itself, creating this intermediate area that remains in a constant state of seasonal vegetative flux.

"See those tracks over there? They are from nutria, the scourge of these marshes," Michael says. "They are large herbivorous rodents that were introduced from South America, but no bienvenue for those rats. They have destroyed hundreds of thousands of acres of freshwater marshes. We didn't see them at the first station because of the higher salinity."

As if to verify his claim, an alligator cruised by slowly to see what we were doing. "You don't see alligators where the salinity is high. He's probably looking for his own nutria cassoulet."

We motored several more miles closer to the mainland to the third station, which was completely different from either of the two previous ones. As we approached the area we scared up hundreds of coot, small, black diving ducks that paddled furiously across the surface of the water to get airborne. They had been feeding on vast mats of submerged aquatic vegetation, called coontail grass, that lives attached to the bottom of this brackish-water environment. We could see blue crabs sculling through the fronds of coontail grass as we passed over them in the airboat. The salinity at this station was an eighth of what you would find in the open ocean.

The surface grass was entirely *Schoenoplectus americanus* (Bulrush), a very common species in brackish marshes. The *Spartina* grasses were nowhere to be seen, and the bulrush had been severely cropped by nutria, leaving large, empty patches of mud spotting the marsh.

The three stations had been set out like a well-thought-out ecology exam. Each station exhibited a different problem, ranging from saltwater intrusion to destruction by the invasive nutria species.

Long before the recent spill, and even before Katrina, coastal scientists were saying that marsh degradation is the real problem that puts New Orleans at risk. They estimate that every mile of healthy marsh can reduce the amount of hurricane-generated storm surges by a foot, so if these marshes had not been "channelized," New Orleans' levees would not have been overtopped and the city flooded. What was needed was a way to understand in detail what was happening to every part of the millions of acres of marsh that flank this coast.

The three sites we had just visited were part of a program that started when Congress passed the Coastal Wetlands Planning Protection and Restoration Project in 1990. Michael explained that the act set aside money to monitor the entire coastal system of southern Louisiana.

"Today, the Coastal Resources Management Investigation System is the only program in the world where we have guaranteed funding to monitor the entire marsh. CRMS was the brainchild of my boss, Greg Streyer. In 2003, he wrote a paper suggesting that if you monitored a hundred marsh sites every month you would have a 95 percent confidence level that you could accurately characterize the vegetation of the marsh. But if you sampled 400 sites for salinity, vegetation, temperature and soil conductivity as well as vegetation, you could not only characterize the health of the marsh, but see a 20 percent improvement in its overall health. Eventually, they compromised and selected 390 sites that are monitored on a monthly basis. Each site receives a report card so you can see both the short- and long-term changes that have occurred since 2005," says Michael.

Louisiana's Governor Bobby Jindal didn't have any of this in mind when he arm wrestled the $360 million out of BP to use Mississippi River mud berms to prevent oil from entering the marshes. It was clear from the beginning that this was a hare-brained scheme. Only a piddling amount of oil was ever contained by the berms. But by the time the spill was capped, more than 17 million cubic yards of sediments had already been dredged, more than $276 million had already been spent, and Governor Jindal had to justify the expenditure. Within weeks, he started saying that the program would continue and the berms' real purpose had always been to rebuild Louisiana's offshore islands to protect her marshes.

Coastal scientists were shocked. For years they had been trying to wrestle money out of government agencies to restore Louisiana's marshes. Now it was being done with money from the private sector—and from BP no less!

Michael explained that there were no guarantees that the project would work exactly as advertised. Most of the sand would wash offshore in succeeding storms, but in doing so, the underlying structure of the islands would be protected. After the storms passed, succeeding waves would supposedly wash the sand back ashore to rebuild the berms.

While there are no guarantees, it is clear that this is probably Louisiana's last and best chance to finally get it right. Governor Jindal is to be applauded. It is an audacious program that never would have happened without the oil spill. It is ironic that this may be the ultimate legacy of the spill. If so, it won't happen a moment too soon. It is only a matter of time before another Katrina knocks on another one of New Orleans' ornate iron grill front doors.

Chapter Thirteen
Transitioning to the Future
The Khyber Pass, Pakistan
October 4, 2010

It was still dark on the steep slopes of the Hindu Kush Mountains. Wisps of fog twisted down off the snowfields to collect on the high roads of the Khyber Pass. This was the route taken by Alexander the Great to reach the plains of India. This was the route taken by Genghis Khan to steal the riches of Persia. This was the route taken by a thousand camel caravans carrying the silks and fine porcelains of China to the Middle East. This was where 16,000 British and Indian troops were slaughtered in 1842. The Hindu Kush Mountains had lived up to their name, "The Killer of Indians."

As the sun rose, two dozen gunmen, wearing the tribal garb of the Pashtun Taliban, came down from the fog-bound Hindu Kush, carrying Kalashnikov rifles. With grim determination, they sprayed bullets at the hundreds of whimsically-painted oil trucks vulnerably positioned near the border crossing. Within minutes, 20 were up in flames and 11 drivers had been killed or wounded.

It was the fourth attack since a U.S. helicopter had killed three Pakistani guards, and Pakistan had retaliated by closing the border to NATO trucks carrying oil to the American troops in Afghanistan. Hundreds of trucks lay defenseless in the crammed parking lots, and hundreds more had been destroyed in just a few days time.

It was no coincidence that a week before the attack 150 marines from Company 1, Third Battalion, Fifth Marines had arrived in the Helmand Province carrying portable solar panels, energy-conserving lights, solar shields that provided shade and electricity to tents, and solar chargers for the communication equipment and computers at their forward base.

The new devices were designed to replace diesel and kerosene, the petroleum-based fuels that would normally have to be driven through the treacherous Khyber Pass. In a press conference held on the same day as the attack, the secretary of the Navy and former ambassador to Saudi Arabia, Ray Mabus, announced that he wanted all Navy and Marine units to be getting 50 percent of their power needs from renewable energy sources by 2020.

With Congress unable to pass an energy bill, this was a reminder that the military can often see the future more clearly than other sectors of the economy and, simply by ordering new technology, can quickly usher in a new era of energy use.

Winston Churchill had a similar vision when he convinced the British Navy to switch from coal to petroleum, helping the allies win World War II on a "flood of oil." But, out of the nine billion barrels of oil the allies used to win the war, eight billion of them came from the United States.

World War II effectively drained America of its most accessible reserves of oil. Despite Nixon's first calls for energy independence in the 60's and Sarah Palin's braying "drill, baby, drill," the United States has simply not had enough oil in the ground to meet its own domestic energy needs since World War II, and every year the amount of oil we need to import has been growing almost exponentially.

A week after the attacks in the Khyber Pass, the United States received another reminder that it is not Presidents and Congress but new technology and visionaries that usually nudge civilizations into a new era of energy use. On October 12th, Google announced that it was going to throw its financial muscle behind a 350-mile-long underwater cable that would form the backbone of a grid carrying power from offshore wind farms to East Coast markets.

This $5 billion clean energy superhighway would stimulate the construction of a dozen offshore wind farms in locations ranging from Virginia to New Jersey, and provide the East Coast with 50 percent of its electrical power needs. The cable would also make it easier for innovators to raise money to build the offshore wind farms and allow the wind-farm operators to produce electricity in an area of the East Coast where the wind was blowing, and then sell it at a profit to another area where the winds were temporarily calm.

Such farsighted moves by both the military and industry can rapidly lower the cost of renewable energy and make it easier for the general public to switch from oil to non-petroleum sources of power. If the government steps in to provide grants to hasten this process, it could prove to be more effective than passing an unpopular carbon tax or initiating a complex market-based system to cap and trade greenhouse emissions.

Yet, on the same day that Google announced it's plan, the Obama Administration lifted the moratorium on deep-water drilling, weeks before the moratorium was scheduled to expire. It was another reminder that despite all the moves to shift to renewables, the world was still firmly ensconced in the age of petroleum, and that oil companies still wielded considerable political power. All the oil companies had to do was point out that the moratorium was causing the loss of 12,000 jobs and that 33 rigs were lying idle and something was bound to happen. After all, it was only two weeks before the mid-term elections.

Inconvenient though it might be, the truth is that, despite

its few, glaring failings, oil has been about the perfect economic energy source. Once you locate it, oil pumps itself to the surface, can be easily transported in pipelines, and refined into multiple, usable products. Because it is naturally liquid, it continues to be about the only major fuel used for transportation. In fact, the entire operation, from getting oil out of the ground, refined and delivered to your gas station requires almost no costs for human intervention. Customers even pump their own gasoline. How many commodities can top those economic advantages?

Compared to coal, America's other traditionally-abundant fuel, oil is relatively clean. The greatest problem with oil is that it takes millions of years to produce and only a few decades to consume. Some economists contend that we have reached peak oil and only have 30 more years worth of reserves. Modern technology may find some extra hidden pockets of oil, but they will probably not be able to extend the period out much beyond 60 years.

Despite the BP spill and the almost-six-month moratorium on offshore drilling, our civilization is just too dependent on oil not to continue pursuing it, and the only places where oil is still available in commercially-viable quantities is in these hazardous deepwater offshore locations.

The truly amazing thing about the BP spill is that it didn't happen sooner. While it was more likely to happen on a BP rig because of the company's corner-cutting corporate culture, it could have happened just as easily on any one of the 4,000 other rigs operating in the Gulf of Mexico.

While we can improve technology and change corporate and regulatory cultures, continued drilling in deeper and more dangerous locations makes it highly likely that we will have more Macondo-like spills.

What is important to remember is that with all oil's faults, we need that genie-in-the-lamp more than ever. We need the genie-in-the-lamp to help us bridge the gap from our present civilization, based on oil, to a civilization run on electrons provided by natural gas, renewables and safer nuclear power plants. We simply can't afford to squander our remaining petroleum on inconsequential wars and consumptive lifestyles. We only have 30 short years to make the transition from petroleum to renewable sources of energy; we better use them very well.

Chapter Fourteen
Tunnell Vision?
February 2, 2010

Oil spills have a way of ruining people's vacations. In mid-December Wes Tunnell was preparing for Christmas when he received a phone call. It was from the head of the team set up to distribute the $20 million set aside by BP to compensate fishermen for income lost because of the spill. Kenneth Feinberg wanted to know when the crab, fish, shrimp and oyster industries would recover from the spill. Could Dr. Tunnell use his wide-ranging knowledge and experience to write up a short paper on the subject? That would be just great. Oh, and by the way, can you get it to us in two weeks?

It was just the kind of challenge Wes Tunnell enjoyed. Unlike most ivory-towered scientists who only feel safe writing for academic journals, the director of Texas A&M's Center for Coastal Studies liked writing papers that would be used to make actual financial and legal decisions. It felt like the real world. He just wished he didn't have to write it over vacation.

Dr. Tunnell had studied the effects of the Ixtoc spill in the early 80's. At the start of the spill, the young marine biologist, who stills bears a striking resemblance to Indiana Jones, remembered thinking that the spill was going to wipe out the entire Gulf and destroy her entire fishery for decades. The seafood harvest did plummet 70 percent in the first year, but by the third year it had popped back up again. Tunnell had been initially baffled by the rapid recovery. But then he discovered that most of the fisheries had already been overfished prior to the spill. During the shut down period, the fish simply had time to recover. It reminded Dr. Tunnell that fish are fecund little creatures that lay millions of eggs; give them half a chance, and they will quickly rebound to their former abundance.

Tunnell plunged into the BP project, only taking a day off for Christmas and half a day off for New Year's. But, by the end of the two-week deadline, he had the 39-page report in hand. The paper was released on February 2nd, and became the scientific underpinnings for Feinberg's determination of settlements from the compensation fund.

The paper reported that the news was much better than anticipated. Dr. Tunnell predicted that most of the fisheries would be fully recovered by 2012; only oysters could take significantly longer, because they had been almost totally eradicated by intentional flooding to prevent the oil from entering the marshes. The report meant that fishermen could only expect to be compensated for three years of lost income.

Reaction to the report was instantaneous. This was not the way scientific papers were supposed to be written. They should only be completed after months, if not years, of study and peer review. Scientists said that the paper was merely opinion; environmentalists said it was too optimistic, or that deep-sea species might recover far more slowly than shallow-water, commercial species. Though fishermen would receive less compensation, most were relieved they could return to their way of life far sooner than they had been originally and erroneously led to believe.

Dr. Tunnell did not dispute any of the criticism. He admitted that it was a first crack at reviewing the evidence, a process necessary in the real world for making day-to-day decisions.

If I had not gone down to Louisiana to see for myself I would have been far more skeptical of the report. But, in November, when I traveled 50 miles out into the Delacroix marshes with

Michael Bey, we saw the many swirls of red drum hunting in the clear waters of the marsh. We saw shrimp boats full of shrimp in the Gulf and bayous, and we saw millions of ducks, geese and shorebirds returning from their summer migrations, well fed and unoiled. We did not see a single drop or sheen of oil. Of course, it did not get BP off the hook but it certainly was a testament to the resilience of Mother Nature; give her half a chance and she will often find a way to rapidly recover. The report highlighted nature's irony, that if you're going to have a major oil spill, the Gulf of Mexico is not the worst place in the world to have it.

Only a few weeks before, BP had signed an agreement with Russia to drill for oil in their fragile Arctic waters. One of the reasons the Russians gave for partnering with the company was because BP had learned so much from cleaning up after their spill in the Gulf. Vladimir Putin explained his thinking by quoting the telling Russian proverb, "One beaten man is worth two unbeaten men." At first, the remark seemed like a gratuitous poke in the eye of BP and the United States.

When you stop to think about it, however, the world did learn a lot from cleaning up the Gulf. But are the hard lessons we learned in the Gulf of Mexico the right lessons to use in the cold, dark, unforgiving waters of the Arctic Circle? The "well from hell" spilled millions of gallons of oil and jeopardized fisheries for three years. Oil from a similar spill in the Arctic could wipe out fisheries for decades and potentially remain in the environment for centuries. It would not be a pretty sight.

Epilogue
The Mystery of the Missing Gas;
Coda to the Tragedy on the Gulf?
February 14, 2011

In June 2010, Dr. John Kessler sailed aboard the Texas A&M research vessel Pisces to collect chemical samples from the BP spill. He was astounded at what he found. The level of methane gas in the ocean was 100,000 times above normal levels. In places it was close to a million times higher than usual. This was not totally unexpected. After all, a bubble of methane gas had caused the original explosion aboard the Deepwater Horizon and slushy formations of methane hydrates had blocked the containment vessel.

Methane is dangerous and ubiquitous stuff. It is a hundred times more potent greenhouse gas than carbon dioxide. It is also widespread, lying frozen below the subsurface of most deep ocean basins. In fact, sub-oceanic methane is believed to be the largest reservoir of carbon on our planet. If the oceans continue to warm as expected, it could start to melt this frozen reservoir of methane that could be the trigger that drives our planet back into an era of runaway global warming. Now a plume of methane gas was hanging below the surface of the Gulf of Mexico, like a huge planetary fart, poised to trigger runaway global warming.

Such things have happened before. Several decades ago, the U.S missile defense system went on high alert, when a reconnaissance satellite picked up what looked like a nuclear explosion in the Indian Ocean. Experts figured that either India or Pakistan had secretly tested a nuclear device. However, most oceanographers think it was probably just a huge bubble of methane that had become dislodged from the warming substrate and exploded on the surface.

When Dr. Kessler visited the site again in August, he found that, to his utter amazement, the gas had disappeared. What had happened? Where had it gone? All he found were lower levels of oxygen and a few species of methane-eating bacteria called methanotrophs.

Now, Dr. Kessler surmises that methanotrophs are common in the Gulf of Mexico. However, usually they just hang out in small numbers waiting for a natural seep of methane to bubble up to the surface. When it does, the opportunistic little creatures reproduce like crazy and use massive amounts of oxygen to metabolize the methane. Then die they out and disappear from the scene as they largely did after the BP spill.

Doesn't this all sound a little familiar? Remember the genies in the bottle, the planktonic calcifiers that saved our planet from global warming by drawing carbon dioxide out of the atmosphere to make petroleum? Well, the methanotrophs are their distant cousins.

If John Kessler is right, methanotrophs are also ubiquitous global genies that can remove methane from the oceans before it rises to the surface to cause global warming. It is ironic that it took the "worst environmental disaster since the flood," to teach us once again that Gaia still has several undiscovered tricks up her sleeve to protect herself from the damage that humans can do to their own precious planet. This then, is the optimistic coda to the tragedy of the "well from hell."

Suggested Further Reading:

The following is a fascinating book that discusses new research on global warming and the role of planktonic calcifiers in cooling the climate:

The Weather Makers; How Man has Changed the Climate and What it Means for Life on Earth. Flannery, Tim. New York, NY: Grove Press, 2005.

The following are two classic books about the early days of the oil industry:

The Seven Sisters; The Great Oil Companies and the World They Shaped. Sampson, Anthony. New York, NY: Viking Press, 1975.

The Prize; The Epic Quest for Oil, Money and Power. Yergin, Daniel. Clearwater, Florida: Touchstone Press, 1991.

There are several more books that cover the more recent history of the oil industry. They are all interesting and are listed in chronological order:

The End of Oil. Roberts, Paul. Boston, MA: Houghton Mifflin. Boston, 2004.

Oil: Money, Politics and Power in the 21st Century. Bower, Tom. New York, NY: Grand Central Publishing, 2009.

The Squeeze; Oil, Money and Greed in the 21st Century. Bower, Tom. Hammersmith, England: HarperPress, 2010.

References:

Chapter 1, Paleozoic Problems

1. The Weather Makers; How Man has Changed the Climate and What it Means for Life on Earth. Flannery, Tim. New York, N.Y.: Grove Press, 2005.

2. "Gulf of Mexico's Deepwater Oil Industry is Built on Pillars of Salt". Voosen, Paul. New York Times, July 28, 2010.

Chapter 2, D'Arcy's Legacy

1. Oil: Money, Politics and Power in the 21st Century. Bower, Tom. New York, NY: Grand Central Publishing, 2009.

2. The End of Oil. Roberts, Paul. Boston, MA: Houghton Mifflin, 2004.

3. The Seven Sisters; The Great Oil Companies and the World They Shaped. Sampson, Anthony. New York, NY: Viking Press, 1975.

4. The Prize; The Epic Quest for Oil, Money and Power. Yergin, Daniel. Clearwater, Florida: Touchstone Press, 1991.

Chapter 3, The Sun King

1. Baron Browne of Madingley, Wikipedia.org. February 4, 2010. http://en.wikipedia.org/wiki/John_Browne,_Baron_Browne_of_Madingley

2. The Squeeze; Oil, Money and Greed in the 21st Century. Bower, Tom. Hammersmith, UK: HarperPress, 2010.

3. "In BP's Record, A History of Boldness and Costly Blunders." Lyall, Sarah. New York Times, July 13, 2010: Business/Energy &

Environment.

4. "Talking Business, BP Ignored the Omens of Disaster."
Nocera, Joe. *New York Times*, June 19, 2010: Business Section.

Chapter 4, New Orleans

1. "Under Mandate to Get Results, Minerals Management
Service Led Way into Deep Water." DeParle, Jason. *New York
Times*, August 7, 2010: Earth & Climate Section

2. "Vision Led to Crazy Horse Find." Shirley, Kathy. *Oil
Explorer Newsletter*, March, 2009.

Chapter 5, "The Kick"

1. "BP Took Risks on Well Job." Hughes, Siobhan. *Wall Street
Journal*, November 9, 2010.

Chapter 6, The Swarm Aboard the Subsea

1. "Another Big Close Call Altered rules." Brown, Robbie. *New
York Times*, August 17, 2010.

2. "Behind Scenes of Gulf Oil Spill, Acrimony and Stress."
Krauss, Clifford. Fountain, Henry. Broder, John M. *New York
Times*, August 27, 2010: U.S. Section.

3. "Oil Spill: BP Had Wrong Diagrams to Close Blowout
Preventer." McClatchy, *Washington Review*, June 17, 2010.

4. "Questions About the Gulf." *New York Times*, August 23,
2010: Opinion Section.

5. WHOI-led Deepwater Horizon Projects. http://www.whoi.
edu. February 4, 2011.

Chapter 7, The Lower Marine Riser Containment Caper

1. Dudley, Bob. Wikipedia. http://www.wikipedia.org/

2. "We're Gonna Be Sorry." Friedman, Thomas L. *New York Times*, July 25, 2010: Opinion Section.

3. "Undersea Lumberjack Gets Saw Snagged in Pipe." Jonsson, Patrick. *Christian Science Monitor*, June 2, 2010.

4. "Tony Hayward Leaving as BP CEO." Weber, Harry B., Associated Press, July 25, 2010.

Chapter 8, Ixtoc and West Falmouth; Two Spills, Two Outcomes

1. "Damage Lives on After '69 Cape Oil Spill." Daly, Beth. *The Boston Globe*, May 21, 2010: Local/Mass Section.

2. "After Oil Spills, Hidden Damage Can Last for Years." Gillis, Justin and Kaufman, Leslie. *New York Times,* July 17, 2010: Science Section.

3. "BP's Tony Hayward Gives Top Kill 48 Hours," Hays, Kristen. Reuters. May 28, 2010.

4. Ixtoc Oil Spill. February 4, 2011. www.wikipedia.org.

5. "BP Prepares to Take New Tack on Leak After 'Top Kill' Fails." Kaufman, Leslie and Krauss, Clifford. *New York Times,* May 29, 2010: U.S. Section

6. "Oil Spill on the Wild Harbor Marsh." Teal, John and Burns, Kathryn. Cape Cod Museum of Natural History blog. http://ccmnh.wordpress.com/

Chapter 9, The Plume Aboard R/V Endeavor

1. "No Sign of Undersea Plume from BP Spill." Baltimore,

Chris. *Scientific American*, September 21, 2010.

2. Cimelli, Rick. Interviewed by William Sargent. Phone Interview. November 17, 2010.

3. Hurricane Alex, Wikipedia. www.wikipedia.org

4. "Oil Dispersants effects remain a Mystery." Khan, Anima. *Los Angeles Times*, September 4, 2010.

5. Oil Spill symposium. Massachusetts Institute of Technology. September 28, 2010.

6. "Science and the Gulf." *New York Times*, September 20, 2010: Opinion Section.

7. Symposium with Chris Reddy. Massachusetts Institute of Technology. November 10, 2010.

8. "Opportunistic bacteria Feasting Slowly on Underwater Oil in Gulf." Voosen, Paul and Winter, Allison. *New York Times*, August 20, 2010.

9. "WHOI Scientists Map and Confirm Origins of Large Underwater Hydrocarbon Plume in Gulf." *WHOI magazine*, August 19, 2010.

Chapter 10, The Flock at Crane's Beach; Ipswich, Massachusetts

1. "Canada's Birds Feel the Impact of BP Oil Spill." *Nature Canada*, E-Newsletter. www.naturecanada.ca/enews_jun10_oilspill-birds.asp.

2. "Oil Travelling up Gulf Food Chain." Robinson, Campbell. New York Times Green blog, November 8, 2010.

3. "Peregrine Falcon" Skoloff, Brian. Associated Press. October 19, 2010.

Chapter 11, The Peninsula

1. "The 'Great Jindini' Aims to Transmogrify Sand Berms into Barrier Shorelines." Bahr, Len. *LA Coastpost*, November 3, 2010.

2. "Oil Cleanup not Over in Louisiana's Bay Jimmy." Fogel, Stephanie. *USA Today*, October 18, 2010.

3. "Dearth of Research Vessels Hampers Oil Spill Science." Gaskill, Melissa. *Nature*, October 13, 2010.

4. "Spill Cleanup Proceeds Amid Mistrust." Robertson, Campbell and Rudolf, John Collins. *New York Times*, November 3, 2010: U.S. Section

5. "Dead Coral Found Near Spill Site." Rudolf, John Collins. *New York Times*, November 5, 2010: Science/Environment Section.

Chapter 12, BP's Surprising Legacy, Delacroix, Lousisiana

1. "The Great Jindini Aims to Transmogrify Sand Berms to barrier Shorelines." Bahr, Len. *LA Coastpost*, November 3, 2010.

2. Bey, Michael. Interviewed by William Sargent. November 10, 2010

3. Gulf Coast Prairies and Marshes, The Nature Conservancy website. www.nature.org

Chapter 13, Transitioning to the Future, Khyber Pass, Pakistan

1. "U.S. Lifts Ban on Deep Water Drilling." Eilperin, Juliet and

Mufson, Steven. *The Washington Post*, October 13, 2010.

2. "Taliban Attacks NATO Trucks in Pakistan." Khan, Zarar. *Air Force Times* and Associated Press, October 6, 2010.

3. Khyber Pass, Wikipedia. www.wikipedia.org

4. "U.S. Military Orders Less Dependency on Fossil Fuels." Rosenthal, Elizabeth. *New York Times*, October 4, 2010: Science/Environment Section.

5. "U.S. Lifts Ban on Deep Water Drilling," Eilperin, Juliet and Mufson, Steven. *Washington Post*, October 13, 2010.

Chapter 14, Tunnell Vision?

1. Schwartz, John and Schrope, Mark. "Report Foresees Quick Gulf of Mexico Recovery From Oil Spill." *New York Times*, February 2, 2011.

2. "On Gulf Oil Spill's Effects, Doing Science with a Deadline." Schwartz, John and Schrope, Mark. *New York Times*, February 3, 2011.

3. "Ixtoc Spill Offers Gulf Coast Hope." Snell, John. Fox News August 17, 2010.

4. "Russia Embraces Risky Offshore Arctic Drilling." *New York Times*, February 15, 2011.

Epilogue

1. "Teeny Janitors Attack Gulf Spill, Then Vanish," Krulwich, Robert. National Public Radio, February 9, 2011.

2. "Russia Embraces Risky offshore Arctic Drilling," Kramer, Andrew E. and Krauss, Clifford, *New York Times*, February 15, 2010.

3. "Gulf Claims Process Under Fire." Associated Press, February 14, 2010.